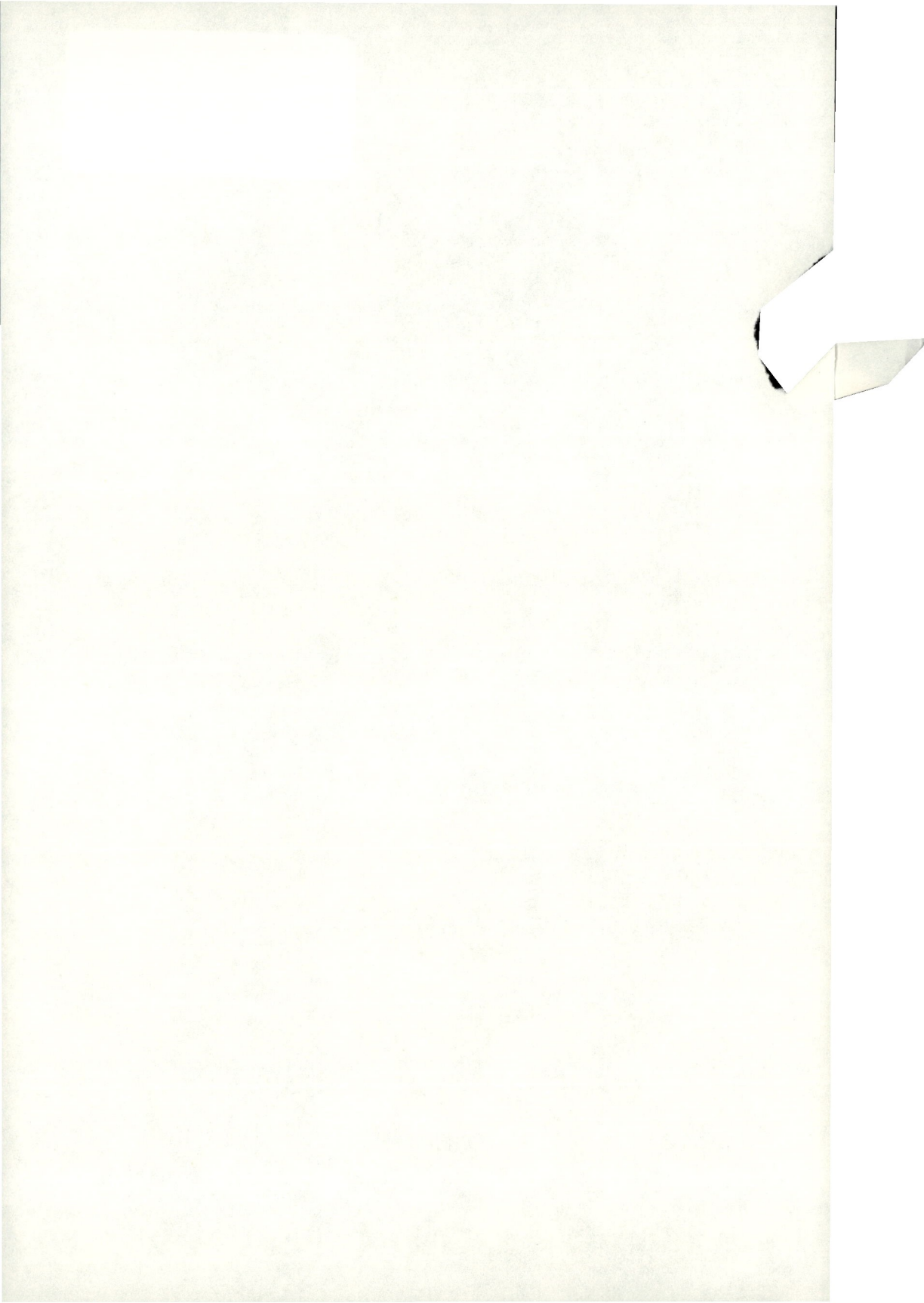

UNIVERSITY OF NEVADA BULLETIN

VOL. XXX	MAY 15, 1936	No. 4

BULLETIN OF NEVADA STATE BUREAU OF MINES AND
MACKAY SCHOOL OF MINES

JOHN A. FULTON, *Director*

———

PLACER MINING IN NEVADA

By WILLIAM O. VANDERBURG
Mining Engineer, U. S. Bureau of Mines

———

PREPARED AND PUBLISHED BY NEVADA STATE BUREAU OF MINES
IN COOPERATION WITH THE UNITED STATES BUREAU OF MINES

———

PRICE 50 CENTS

PUBLISHED QUARTERLY
BY THE
UNIVERSITY OF NEVADA
RENO, NEVADA

1[69]

CARSON CITY, NEVADA
STATE PRINTING OFFICE   -   -   JOE FARNSWORTH, SUPERINTENDENT
1 9 3 6

TABLE OF CONTENTS

LIST OF ILLUSTRATIONS

LIST OF TABLES

FOREWORD

On December 15, 1932, a bulletin on placer mining in Nevada was issued by the Nevada State Bureau of Mines in cooperation with the United States Bureau of Mines; this issue was exhausted some months ago, but the demand for it has increased.

The price of gold was raised to $35 per ounce by Executive Order in January, 1934, whereas the price was $20.67 per ounce in 1932 when the first placer bulletin was issued. The higher price of gold has stimulated interest in placer mining greatly and has created an urgent demand for reliable information on placer deposits in this State, and methods of working them.

To meet this situation, the Director of the Nevada State Bureau of Mines consulted with Dr. John W. Finch, Director of the United States Bureau of Mines, with the result that a cooperative arrangement was entered into by the two bureaus, and William O. Vanderburg was detailed by the Federal Bureau to make a survey of the placer mining industry of the State and to report on it. The present bulletin was written as a result of this cooperative arrangement.

JOHN A. FULTON,
Director Nevada State Bureau of Mines.

PLACER MINING IN NEVADA[1]

By WILLIAM O. VANDERBURG[2]

INTRODUCTION

Placer mining has always been an attractive and often profitable branch of the mining industry. A large part of the world's gold supply has been obtained from the earth by placer mining methods. The search for placer gold and the exploitation of the deposits when found has been an important factor in the development of our western States. Since the dawn of history gold has been preeminent among the metals and the objects of human desire. The beneficent part gold has played in the development of civilization, as a standard of value, and as a medium of exchange, exceeds all reckoning.

Gold mining is a depression-proof industry, and unlike other commodities there is always a ready market for the product no matter what the production may be. The gold mining industry is to a great extent immune to the hazards which affect other industries, as the sale of the product is unaffected by overproduction or competition.

In 1935, the writer estimates that about 500 men were employed in placer mining within the State. In recent years there has been a marked tendency to employ mechanical equipment in the exploitation of placer deposits in Nevada. In 1932, when the first survey of placer mining was made in the State, most of the operations were carried on by hand methods. In 1935, at least 15 placer operations employed tractors and scrapers, drag lines, power shovels and trucks for mining, and trommel screens in conjunction with various types of gold saving equipment for treating the auriferous gravels. In 1935, a number of small hand operations were scattered over various placer deposits throughout the State. Most of the men who work with dry washers, rockers, and like equipment are nomadic in their pursuits and go to the camps only when weather conditions are favorable.

The revival of placer mining no doubt will have its unfavorable features, as some of the workers who do not fully understand placer-mining methods and conditions will find themselves worse

[1]Published by permission of the Director of the U. S. Bureau of Mines. (Not subject to copyright.)
[2]Mining Engineer, U. S. Bureau of Mines. Manuscript completed February 29, 1936.

off at the end than at the beginning. The general results, however, cannot fail to be beneficial and are apparent in the increased yield from the placer deposits of the State.

Ability to work the placer deposits in the State on a scale larger than could be done by hand methods depends on the question of water supply, efficient operation, and a thorough understanding of the problems involved. Many failures have resulted and will result from attempts to use special methods and freak equipment. Many such have been tested and their inefficiency demonstrated, whereas, with proper equipment, success might have been achieved.

In addition to the information gathered by the writer in the preparation of this report, placer data that appeared in publications of the United States Geological Survey, Nevada State Bureau of Mines, and in the technical press have been used freely.

Thanks are due Director John A. Fulton, Mackay School of Mines, University of Nevada, for assistance in the preparation of this bulletin. The writer also wishes to acknowledge his indebtedness to the various placer operators within the State, who generously supplied information without which this bulletin could not have been prepared.

SCOPE OF REPORT

In the past, lode mining within Nevada has so overshadowed placer mining that there is a lack of published data on the State's placer deposits.

In 1932 "Placer Mining in Nevada," was prepared by Alfred Merritt Smith and the writer and published by the Nevada State Bureau of Mines. That bulletin was so favorably received by prospectors and others interested in the development of placer deposits that it was soon out of print and it was deemed advisable to bring it up-to-date in a new edition. This bulletin is not intended as an exhaustive report, but merely as a summary of the present status of placer gold operations.

In collecting data for the present bulletin the writer spent from October 31 to November 30, 1935, in making a field survey, during which he traveled about 5,000 miles by automobile and visited 37 placer districts. Due to bad weather and lack of time it was impossible to visit several placer districts in the northern part of the State.

Most placer deposits in the State occur in arid or semiarid regions where the water available is insufficient to exploit the deposits by orthodox placer-mining methods. The methods for recovering gold from the dry placers, the location of the deposits, equipment used, and similar matters, have been the subject of many inquiries received by the State Bureau of Mines. The writer trusts that these inquiries have been answered in the following pages in greater detail than would be possible by correspondence.

BRIEF HISTORY OF PLACER MINING

The term "placer" is derived from the Spanish and originally signified a gravelly place in a stream bed or by the side of a stream where gold occurred. In the United States the word is applied to all detrital deposits containing valuable minerals, such as gold, platinum, tin, cinnabar, or tungsten, having their origin in flowing water and possessing characteristics of a gravelly nature. An exception should be made of residual placer deposits, for streams have nothing to do with their origin.

The earliest placer mining appears to have been carried on by Paleolithic man while searching the river beds for flints from which to manufacture his stone implements. Evidence of prehistoric mining for flints has been discovered in the south of England and in the Somme Valley in France.

Placer gold deposits were, perhaps, the first deposits of metals to be mined because they yielded gold free from base metals or gangue. References to alluvial mining date back to the dawn of civilization.

The earliest known records of placer mining for gold are Egyptian. A little gold has been found in the pre-dynastic Egyptian graves belonging to a period of about 6,000 years ago. Egyptian hieroglyphics from the period of 3400 B.C. indicate that placer gold in the form of electrum was derived from placers.

Figure 1 is a reproduction of a papyrus map in the archives of the University of Turin. This map was made about 1300 B.C. and it is the oldest map known. It portrays among other things "mountains in which gold is washed," "houses of the gold working settlement," and "mountains of gold."

Pliny (Caius Plinius Secundus), who was appointed Procurator in Spain in 66 A.D., has given an interesting though somewhat confused account of Roman placer-mining operations in Spain in his "Natural History." According to Pliny, "Gold is found in the world in three ways, to say nothing of that found in India by the ants and in Scythia by the Griffins. The first is as gold dust found in streams, as, for instance, in the Tagus in Spain, in the Padus in Italy, in the Hebrus in Thracia, in the Pactolus in Asia, and in the Ganges in India; indeed, there is no gold found more perfect than this as the current polishes it thoroughly by attrition. * * * Others by equal labor and greater expense bring rivers from the mountain heights, often for a hundred miles, for the purpose of washing this debris. The ditches thus made * * * entail a thousand fresh labors. The fall must be steep, that the water may rush down from the very high places, rather than flow gently. The ditches across

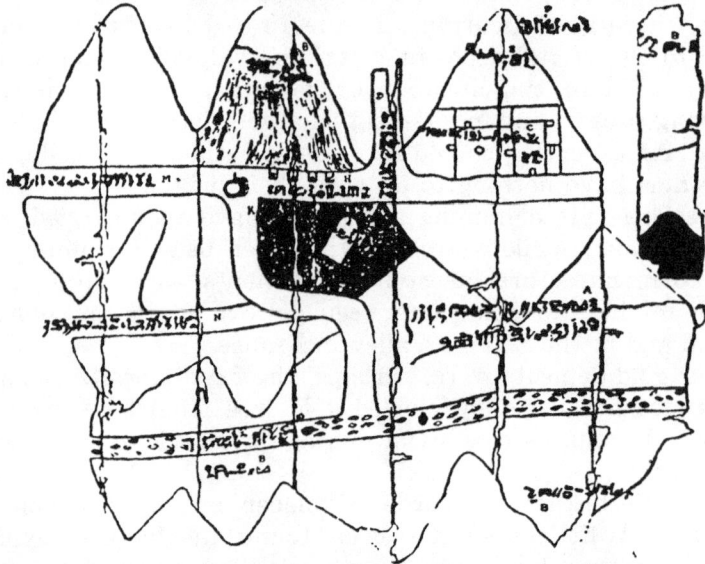

Figure 1. Papyrus map of an ancient Egyptian gold mining district
(About 1300 B.C.)

the valleys are joined by aqueducts, and in other places, impassable rocks have to be cut away and forced to make room for troughs of hollowed-out logs. Those who cut the rocks are suspended by ropes, so that to those who watch them from a distance, the workmen seem not so much beasts as birds. Hanging thus, they take the levels and trace the lines which the ditch is to take; and thus, where there is no place for man's footstep, streams are dragged by men * * *.

"When they have reached the head of the fall, at the top of the mountain, reservoirs are excavated a couple of hundred feet long and wide, and about 10 feet deep. In these reservoirs there are generally five gates left, about 3 feet square, so that when the reservoir is full, the gates are opened, and the torrent bursts forth with such violence that the rocks are hurled along. When they have reached the plain there is yet more labor. Trenches *. * * are dug for the flow of water. The bottoms of these are spread at regular intervals with ulex to catch the gold. The ulex is similar to rosemary, rough and prickly. The sides too,

—SLUICE. B—BOX. C—BOTTOM OF INVERTED BOX. D—OPEN PART OF IT. E—IRON HOE. F—RIFFLES. G—SMALL LAUNDER. H—BOWL WITH WHICH SETTLINGS ARE TAKI AWAY. I—BLACK BOWL IN WHICH THEY ARE WASHED.

Figure 2. Long tom and prospector's pan used in medieval Europe. (After Agricola.)

are closed in with planks and are suspended when crossing precipitous spots. The earth is carried to the sea and thus the shattered mountain is washed away and scattered, and this deposition of earth in the sea has extended the shore of Spain. The gold procured * * * does not require to be melted but is already pure gold. It is found in lumps, in shafts as well, sometimes even exceeding ten Roman pounds in weight. * * * The ulex is dried and burned and the ashes are washed on a bed of grassy turf in order that the gold may settle thereon."

Thus it is seen that the Romans practiced booming and sluicing of placer ground in Spain before the Christian era.

Agricola,[3] writing about 1556, described the equipment used at that time for placer mining, which included pans, rockers, and

A—STREAM.   B—DITCH.   C—MATTOCK.   D—PIECES OF TURF. E—SEVEN-PRONGED FORK.
F—IRON SHOVEL. G—TROUGH. H—ANOTHER TROUGH BELOW IT. I—SMALL WOODEN TROWEL.

Figure 3.   Ground sluicing in medieval Europe.   (After Agricola.)

sluices, the latter both with and without riffles.   Figures 2 and 3 illustrate several methods of placer mining in medieval Europe.

[3]Agricola, Georgius. *De Re Metallica.* Translated from the first Latin edition of 1556 by Herbert Clark Hoover and Lou Henry Hoover, 1912, pp. 321–348.

When riffles were used, the bottoms of the sluices were covered with cloth, turf, woven horsehair and the like to catch the gold. Agricola mentions that the Colchians placed skins of animals in water to catch the gold. The legend of Jason and the Golden Fleece, which captures the imagination of the schoolboy, may have had its foundation in a piratical expedition for the purpose of looting the placer mines in Asia Minor, where fleeces were employed to catch the gold in much the same manner that cocoa matting, burlap, and corduroy are used today.

The foregoing is sufficient to indicate that the simple equipment used by the present-day placer miner was well known to the ancients. In fact, some of the devices employed today have come down the centuries with very little improvement.

Hydraulic mining is a comparatively recent development, having been introduced into California by miners from Dahlonega, Ga., about 1850. At first, water under pressure was ejected against the gravel bank through cotton hose with nozzles attached, similar to our present fire-fighting apparatus. Subsequently, iron pipe was used instead of hose, and the nozzle was changed in shape and increased in size until the hydraulic giant or monitor was evolved. With the development of the hydraulic giant, placer mining expanded rapidly, until, in 1876, in California alone there was over $100,000,000 invested in flumes, ditches, and other equipment for hydraulic mining.

In California the debris from hydraulic operations was so great that it overflowed on agricultural lands and filled in navigable river channels, so that a number of hydraulic operations, not having means to store their debris, were forced to cease operations by the injunction decision of Judge L. B. Sawyer in the U. S. Circuit Court in 1884. The Caminetti Act, passed by Congress in 1893, placed hydraulic mining on the drainage systems of the San Joaquin and Sacramento Rivers under the supervision of the California Debris Commission.

The use of dredges for working placer ground is the most recent development in placer mining. The first successful dredge, forerunner of the modern and efficient dredges, was operated in New Zealand in 1866. Gold dredging in the United States began about 1898.

There is some evidence that placer deposits were worked in Nevada prior to 1848, while it was still a part of Mexico. The Tule Canyon placers are said to have been operated in the early days by Mexicans, who recovered over one million dollars. The evidence is rather vague and the production of this amount strains one's credulity.

The first authentic discovery of placer gold in the State was made by Abner Blackburn in July, 1849, in Gold Canyon below Virginia City. This first discovery did not cause any great excitement. From 1849 to 1857 as many as 200 men were employed in the placer diggings in Gold Canyon. In 1857, a number of men from Gold Canyon prospected in Six-Mile Canyon, the next ravine two miles north of Gold Canyon. Here the gold was found not in auriferous sand and gravel, but in a blue clay that had to be disintegrated in water to free the metal. From $5 to $13.50, the value of an ounce of the metal, was a day's wages. The miners were puzzled to understand the peculiar characteristics of the "placer" gold. Working up both ravines, the placer gold miners eventually struck rich "pay" in the decomposed outcroppings of the prominent quartz ledges crossing the head of both ravines. In the fall of 1859, the miners began to understand why the gold they had been washing out of the alluvium below the big quartz ledges was only worth about half as much as the gold found in the placers on the western slope of the Sierra Nevada; the decomposed quartz, almost black in color, was found to be rich in silver in combination with the gold, and the famous "Washoe" excitement followed. This was the beginning of mining on the Comstock Lode and the discovery of the first noted silver mining district in the United States.

In the hegira of '49 to the gold diggings of California the early pioneers, lured by the pot of gold at the end of the rainbow, hurried over the arid stretches of Nevada with a minimum of delay. On reaching the gold fields the disappointment that awaited many caused the backwash of this first tide of gold seekers to spread over the territory east of the Sierras to prospect for other sources of mineral wealth. Many of these came to Nevada, and in the early days wave after wave of prospectors were led from place to place by stories of fabulous riches. As a result of this prospecting, many placer deposits as well as lode deposits were found and worked. From the early days to the present time prospecting has occurred in cycles; whenever a strike of major importance was made it acted as a stimulus to renewed activity only to die down after the first excitement had subsided, to burst forth anew when another strike was made.

Mention should be made of the important role taken by the Chinese in developing the placer deposits in the State in the early days. Chinese labor was first introduced in Nevada in

1858 to work on the placer ditch that Orson Hyde began and J. H. Rose completed to take water from the Carson River to the mouth of Gold Canyon. Large numbers of Chinese were later employed in grading roadbeds while the Virginia and Truckee Railroad and the Central Pacific Railroad were being built, the latter being completed across the State on May 10, 1869. The Chinese never were welcomed to Nevada and were discriminated against in the laws and the constitution of the State, their employment being prohibited also by the charters of the railroads constructed within the State after 1871. In spite of legislative enactments and popular feeling, it was impossible to prevent them from engaging in placer mining, and after the railroads were completed large numbers of Chinese took up

Figure 4. Chinese placer miner with rocker and other tools.

placer mining, spreading over the northern part of the State. Placer mining by the Chinese was found to be profitable, and a large number were brought from their homeland by their fellow countrymen for the express purpose of conducting placer mining. A typical Chinese placer miner with his equipment is shown in Figure 4. The majority of the Chinese placer miners withdrew from the State prior to 1900.

In recent years placer mining activity within the State has been revived to a considerable extent. Today, on the routes formerly traversed by ox teams and covered wagons and later by the pony-express riders, itinerant miners and tourists travel over well - maintained highways at speeds limited only by the capability of the machines or the caution of the drivers.

PRODUCTION OF PLACER GOLD IN NEVADA

The amount of gold recovered from Nevada placers prior to 1900 will never be known with accuracy as no complete authentic records are extant. For the placer production prior to 1900, one can only be guided by information gathered from old-time residents familiar with the activities of the various districts over a long period and from estimates made in early reports. A number of the placer deposits in the State discovered prior to 1900 were first worked by Americans, who skimmed the deposits of their richer values, and later by Chinese, who were content with smaller returns. The Chinese were secretive as to their earnings, and much of the gold produced by them neither found its way to the mints nor was handled by the Wells Fargo Express Company. In many parts of the State signs of former placer activity may be seen in the form of numerous filled-in shafts sunk in gullies and canyons and old tailings piles partially obscured by growths of sagebrush.

Prior to 1900, American Canyon and Spring Valley in the Humboldt Range are known to have been heavy producers, and their production is estimated at $10,000,000. Barber and Wright Canyons in the Sierra district are each reported to have produced $2,000,000. The Osceola district is said to have produced between $2,000,000 and $3,500,000. The placers in the vicinity of Tuscarora are known to have produced in the neighborhood of $7,000,000. Several deposits in northern Elko County in the vicinity of Charleston are also known to have attained a considerable production of placer gold in the early days. Gold Canyon and Six-Mile Canyon, below Virginia City, produced considerable placer gold both before and after the discovery of the rich lode mines of the Comstock.

Allowing for the tendency to overestimate the production of the districts worked in the early days and for the large number of placer districts that were worked on a small scale with no record of production, it is estimated that the total production of placer gold in Nevada from 1849 to 1900 was about $25,000,000. It has been pointed out to the writer by old-time placer miners that in some cases the production has been more than stated in this bulletin. Probably it has been, but as there are no recorded figures to substantiate the claims it is best to err on the side of conservatism.

The production of placer gold and silver from 1900 to 1935, as compiled from the annual Reports on Mineral Resources of the United States published by the United States Bureau of Mines is shown in the following table:

TABLE 1
Annual Production of Placer Gold and Silver in Nevada from 1900 to 1935

Year	Gold (fine ounces)	Value	Silver (fine ounces)	Value	Total value
1900	990.0	$20,465	....	....	$20,465
1901	1,621.0	33,509	....	....	33,509
1902	757.0	15,649	....	....	15,649
1903	1,762.0	36,424	....	....	36,424
1904	1,460.5	30,191	....	....	30,191
1905	400.2	8,273	98	$59	8,332
1906	2,556.1	52,839	1,296	866	53,705
1907	2,673.9	55,274	1,846	1,206	56,480
1908	3,857.9	79,750	2,013	1,064	80,814
1909	4,013.4	82,964	1,534	790	83,754
1910	7,854.7	162,371	3,655	1,955	164,326
1911	10,181.1	210,462	4,536	2,418	212,880
1912	11,206.2	231,652	4,474	2,724	234,376
1913	14,775.8	305,442	6,363	3,814	309,256
1914	18,250.1	377,262	6,367	3,490	380,752
1915	19,123.6	395,319	6,030	2,996	398,315
1916	17,139.9	354,313	5,744	3,772	358,085
1917	14,153.8	292,584	6,278	5,112	297,696
1918	10,564.1	218,380	3,907	3,907	222,287
1919	6,399.4	132,288	2,374	2,659	134,947
1920	7,383.9	152,639	3,296	3,593	156,232
1921	17,567.0	363,142	7,956	7,956	371,098
1922	11,602.4	239,842	4,710	4,710	244,552
1923	3,941.8	81,484	1,783	1,462	82,946
1924	1,324.0	27,369	546	366	27,735
1925	2,536.5	52,435	1,285	892	53,327
1926	2,866.2	59,249	1,494	932	60,181
1927	1,809.2	37,400	740	419	37,819
1928	1,851.1	38,266	839	491	38,757
1929	2,117.0	43,762	1,025	546	44,308
1930	1,859.4	38,437	847	326	38,763
1931	2,883.2	59,602	860	249	59,851
1932	5,408.2	111,798	1,743	492	112,290
1933	5,769.5	119,267	1,991	697	119,964
1934	5,248.9	183,449	1,594	1,030	184,479
1935	6,600.0	230,670	2,000	1,476	232,146
1935	230,509.0	$4,934,222	89,224	$62,469	$4,996,691

NOTE—Production for 1935 subject to revision.

From 1900 to date the bulk of the production was from the Manhattan, Round Mountain, and Battle Mountain placers. The production of placer gold from Manhattan from 1907 to 1930 was $1,325,173; from Round Mountain from 1906 to 1932, has been $1,295,920; and from the Battle Mountain placers from 1910 to 1935, approximately $900,000. The placers at Rawhide are reported to have yielded about $250,000.

TOPOGRAPHY AND HYDROGRAPHY OF NEVADA

Nevada, with its 110,690 square miles of thinly populated territory includes the larger portion of that great undrained area known since the explorations of Fremont as the Great Basin. On the west the Great Basin is bordered by the Sierra Nevada Range and on the east by the Wasatch Range. The country in between these two ranges bears little resemblance to a basin, however. A relief map of Nevada shows a series of nearly parallel mountain ranges trending northerly and southerly, separated by troughlike valleys. This alternation of valleys and mountain ranges is preserved with great uniformity throughout the State. The elevation of the valleys varies from 4,000 feet above sea level in the northern part to about 2,000 feet in the southern part. The mountains range from 5,000 to nearly 12,000 feet above sea level. Above a general elevation of 7,000 feet, the mountains are usually covered with sparse growths of pinion pine, mountain mahogany, and other scrubby trees. Below this elevation throughout the entire State the wild sage or artemisia grows, but for a few short weeks in the spring of the year the desert is alive with flowers, which relieve the monotony of the sagebrush gray.

The system of mountains and valleys forms a series of basins resulting in a unique hydrography, the receptacles of the drainage waters becoming, according to conditions, lakes, sinks, alkali flats, or salt beds. In former geologic time many of the valleys formed large lakes, remnants of which are still present in Walker, Pyramid, and Ruby Lakes, each having a drainage system of its own.

Nevada is considered to embrace the most arid portion of our United States, but the water resources are greater than they are generally supposed to be. The average annual precipitation in Nevada for the period 1915 to 1926 was 8.39 inches. This precipitation is not uniform, and in general it increases in proportion to the elevation, with the mountains receiving more and the valleys less than the average. Most of the precipitation is in the form of snow during the winter months.

The sides of the mountain ranges are traversed by numerous canyons, and in these canyons, during the spring of the year, are small streams formed by melting snow. These seasonal streams flow into valleys filled with loose sediments which form favorable underground storage basins for the accumulation of water. Evidence of the presence of these underground reservoirs is found in springs, seeps, marsh land, and in the numerous drilled wells which furnish a large portion of the water supply for towns and communities within the State. The depth of the sediments in the valleys is very great; drilling records showing depths up to 2,000 feet in certain localities. These deep sediments store large quantities of water and in places where conditions are favorable for artesian flow, the yield is correspondingly large.

The only meteoric waters in the State that reach the ocean by surface flow are a few small streams tributary to the Snake River in the northern part of the State and those tributary to the Colorado in the southern part. With these exceptions, the other rivers, including Truckee, Walker, Carson, Humboldt, and numerous creeks, are collected in drainage areas within the State. A large portion of the smaller rivers within Nevada may be said to flow "upside down"; that is, most of the waters sink beneath the surface and travel along bedrock. Such underflow waters play an important part in the development of water resources for placer mining and other purposes.

The character of vegetation is often indicative of underground water; rabbit brush, salt grass, and the false goldenrod indicate a relatively high water table. Greasewood, which grows over a wider range than the plants mentioned, also indicates a high water table. In some cases the greasewood taproots go down nearly 50 feet in search of water.

GENERAL GEOLOGY OF NEVADA PLACERS
DERIVATION OF PLACERS

Gold placer deposits have been derived from the disintegration and weathering of auriferous veins and mineralized rocks. The disintegration of rocks is accomplished slowly by natural agents, namely, the wind, rain, flowing streams, frost, changes in temperature, growth of vegetation, chemical action, and movements of the earth's crust. These agents, working throughout geologic time, reduce the rocks to gravel, sand, silt, and clay, and liberate the gold. The gold may have existed in native form alloyed with silver or in association with other minerals. The usual occurrence of gold is as native metal associated with quartz in veins or small seams in country rock. Running water transports the loosened rock and much of the gold, if conditions are favorable, away from its place of origin. Since gold has a high specific gravity, about seven times that of quartz, with the ratio increasing to eleven times under water, it tends to work down to the bottom of the moving material and is concentrated on the bedrock or on impervious strata of clay known to miners as "false bedrock."

Usually gold placer deposits are found in districts where lode-gold deposits occur, although the original source may not have been a deposit which could be mined profitably. Where the deposits originally existed at shallow depths, erosion may have been so complete that not even the roots of the veins can be found.

RECONCENTRATIONS

Placers may be formed solely by rock weathering with little movement of the gravel, but more commonly they are formed by water transportation, sorting, and deposition. Many of the richest placers are those formed by the erosion of older placers and the reconcentration of their gold. The second movement of the gold may have been caused by a change in the gradient of the stream, due to rising or subsidence of the earth's crust, or to an increase in the flow of water at a confluence of streams, which would cause increased velocity to sweep away old gravel deposits.

ACTION OF WATER

Geologic evidence shows that there was an abundance of water in the desert regions of Nevada at one time. This is proven by the presence of fossil shells and lacustrine deposits in isolated patches throughout a large portion of the State. The evidence of water action in the formation of gold placers is shown in some of them by the well-rounded pebbles and gold particles. The placer material was transported by gravity, rains, freshets and melting snows into the beds of ravines or streams that are now dry. The coarsest gold is usually found at the heads of the ravines or canyons, while the fine gold may have been borne many miles from its source.

Nearly all channels have barren as well as rich spots, due to conditions prevailing at the time of the formation of the channel and the deposition of the gravel and gold. Stream beds are seldom rich directly opposite the mouth of a tributary stream, though both streams may contain gold, for at that point the increased volume and velocity of water will have swept the gold onward, to settle at some point where the current is slower. Neither will much gold be found in rapids or whirlpools. Whirlpool pits may seem ideal spots for the concentration of gold, but the gold that lodges in such holes is usually thrown out or ground to powder by the attrition of milling gravel and boulders. Repeated exploration of such places has resulted too often in disappointment.

Gold often is found in loose alluvium or wash, in which there are no rounded pebbles or rocks. This is particularly so in the desert areas of Nevada, where disintegration of the rocks produces a mass of loose material, mostly angular, which is washed to lower land by the infrequent but heavy rains that occur in the State. Much of the finer material may be blown away by the winds during the slow process of erosion and concentration.

The gold occurring in such deposits is rough or angular and shows little evidence of abrasion or rounding.

In the more arid portions of Nevada, placer-gold concentrations may be erratic, due to cloudburst action. Rainfalls on the desert are usually short-lived but violent, and the greatest force is always in the mountains, where there is little soil or vegetable growth to retard the floods. When a cloudburst occurs, a great torrent of water, sand, and gravel pours down the canyons to the lowlands. The great onrush of water destroys the slowly formed concentrations of gold by washing them on to the deltas, so that the gold in the alluvial fans usually is erratic in its distribution. Coarser gold often is found mixed with sand and gravel in crevices in bedrock. Placers containing fine gold in surface wash often have the gold associated with or imbedded in clay. Sand and gravel in water become more or less mobile, so that the coherence between the grains is diminished, with the result that any fine gold particles in the mass tend to slip down between the interstices of the sand and gravel and settle on more impervious clay formations beneath. Clay deposited by a stream becomes coherent and plastic and the fine gold is held by it. Clay also works under boulders in the stream beds and is protected there from the disintegrating action of the water. Fine gold, moving along over the bottom of a stream channel, works its way beneath the boulders and becomes imbedded in the clay, adhering to the under side of the boulders; very often rich pockets of gold are found in clay accumulations under large boulders close to but not on the surface of the bedrock.

In the more arid parts of the State, placer gold was deposited largely by intermittent rather than constant stream action. When cloudbursts occur in the arid regions, the flow of water lasts only a few hours; there is no time to soften the underlying earth except for a few inches and, in consequence, the gold is found in narrow streaks of limited length and depth. The pay streaks are very irregular in occurrence and have to be located largely by hit-or-miss prospecting. Rules of prospecting derived from a region of constant streams are likely to prove more of a hindrance than a help in prospecting dry placers.

In dry placers we find the pay streaks sometimes at or near the true bedrock, but more often at various higher levels, usually resting on layers of caliche* that form a series of false

---

*Caliche is a word used in Mexico and in the southwest United States for gravel, sand, or desert debris cemented by porous calcium carbonate; also the calcium carbonate itself.

bedrocks. The gravels in the dry placers are usually a mixture of fine and coarse material showing little evidence of stratification.

TYPES OF PLACERS

Based on their mode of origin, placers have been classified by Jenkins [4] into seven types, as follows:

1. Stream placers (alluvial deposits), sorted and resorted, simple and coalescing.

2. Eluvial or hillside placers, representing transitional creep.

3. Residual placers or "seam diggings."

4. Bajada placers, a name applied to a certain peculiar type of "desert" or "dry" placer.

5. Glacial - stream placers, gravel deposits transitional from moraines.

6. Eolian placers or local concentrations caused by the removal of lighter materials by the wind.

7. Marine or beach placers.

Placers are sometimes classified according to their topographical position, such as bench, river bar, gulch, gravel plain, and others.

In the genetic classification of placers there is no sharp line of demarcation between one type and another, as two or more fundamental processes of deposition may have been involved in their formation.

In Nevada the stream placers are the most important. Sorted stream placers are those in which the gold has been freed from its matrix and transported by water to a new location. Very often gold is found in gulches and ravines now dry or containing very little water. Some stream deposits may be found on the tops and sides of hills where they were left by streams that changed directions or disappeared entirely as the surface of the earth changed. Streams often deposit gold at bends where the velocity of the transporting water decreased, thereby forming bars and allowing the gold to settle. As the streams cut deeper and the direction of flow changes, such bars are often left behind, forming bench or terrace placers, so-called because they are at a higher elevation than the present level of the streams.

Eluvial or hillside placers are formed by the disintegration of lodes on a hillside. The gold, along with the disintegrated country rocks, may gravitate or creep slowly down the hill assisted by natural agents such as frost or trickling rain

---

[4]Jenkins, Olaf P., New Technique Applicable to the Study of Placers, California Journal of Mines and Geology, vol. 31, April, 1935, p. 156.

waters, thereby forming gold concentrations of economic importance. A number of placer deposits in the State, such as those
at Round Mountain and on the west slope of the Snake Range
in the Osceola District, were formed partly by this process.

Residual placers are those in which the gold has accumulated
in place by the disintegration of its bedrock matrix. The gold
has not moved far from its source. The Round Mountain deposits are partly of this type. It was while working such a placer
deposit at Round Mountain that a prominent vein was discovered.

The bajada type of placer has been described by Webber.[5]
Bajada is a Spanish word meaning slope and it is used in the
Southwest to indicate the detrital material on the lower slopes
of a mountain range. The genesis of a bajada placer is stated
by Webber to be basically similar to that of a stream placer
except as it is conditioned by the climate and topography of the
arid region in which the placer occurs.

Glacial stream placers, as their name implies, are formed by
the action of glaciers. In early geologic time glaciers occurred
in the high Sierra. Some engineers believe that the Mount Siegel
placer deposits in the Pine Nut Range in the northwestern part
of the State were formed in part by glacier action.

Eolian placers are those wherein the gold is concentrated at
the surface by the action of the wind in removing lighter material. Eolian placers have been worked in the arid regions of
Australia, but they are of little economic importance in Nevada.

Beach placers are formed by the action of shore waves and
currents. Concentrations of gold and other heavy minerals by
wave action have been worked along the Pacific Coast and
Alaska.

ANCIENT PLACERS

In several localities of the State placers occur that were formed
by ancient streams of Tertiary age. No doubt some of these
ancient channels contain gold and are of economic importance,
and a careful geologic study may lead to the discovery of ancient
placers in places within the State that heretofore have escaped
observation. The old river channels of the Sierra Nevada in
California have contributed many millions of dollars to the
world's gold supply and they are still productive. A number
of these old channels are fairly well preserved on the eastern
slope of the Sierra Nevada in the Carson and Markleeville quad-
rangles. A particularly large and well-defined ancient stream

---

[5]Webber, Benjamin N., Bajada Placers of the Arid Southwest, Amer. Inst. of
Min. and Met. Eng., Tech. Publ. 588, 1935.

has been described by Reid.[6] This Tertiary river had an east-west course, and has been traced from the northeast corner of Lake Tahoe to a point about four miles north of Carson City. This river is believed to have flowed eastward, although complete evidence is lacking. To the east the gravels disappear in the Virginia Range.

At Genoa, 15 miles south of Little Valley, placer gold is found in well-washed gravels believed to be of similar Tertiary origin. East of Genoa in the Mount Siegel District, Tertiary placer gravels are found at a considerable elevation. Several miles farther east are the recently discovered placers of Smith Valley and Yerington, some of which are probably of Tertiary origin.

In the Charleston and Mountain City Districts of northern Elko County are large areas of Tertiary gravel overlying rhyolite and sedimentary rocks. Concentration of these gravels by both ancient and recent streams has resulted in the formation of placers of economic importance. About 15 miles north of Jarbidge, on the east fork of the Jarbidge River, the deep canyon exposes successive flows of rhyolite bedded on each other. Over these rocks lie beds of clay and well-washed rounded pebbles, which are capped by a recent basalt flow (figure 5). These gravels are similar to those of Tertiary age in the Charleston and Mountain City Districts.

In Walker Gulch, Rochester District, Pershing County, gravels occur which, according to Schrader,[7] are of Tertiary age. After having been deposited by an ancient stream, the gravels were covered by several thousand feet of basalt. The basalt has now been worn away in places, exposing the gravels. In 1913 and 1914 some placer gold was recovered from the gravels by drift mining. Southwest of Walker Gulch, in Limerick Canyon, terrace gravels have been mined profitably. These gravels are of interest as they may be parts of an ancient river channel that crossed the Humboldt Range westerly through Spring Valley Pass, and the gravels, dispersed by later erosion, enriched the placers found in Spring Valley, American, and South American Canyons.

RELATION OF PLACERS TO LODE DEPOSITS

Sometimes placer deposits have led to the discovery of important lode mines, but in many cases the lode sources of the placer

---

[6]Reid, John A., A Tertiary River Channel near Carson City, Nevada: Min. and Sci. Press, vol. 96, 1908, pp. 522–525.

[7]Schrader, F. C., The Rochester Mining District, Nevada: U. S. Geol. Survey Bull. 580, 1913, pp. 368–370.

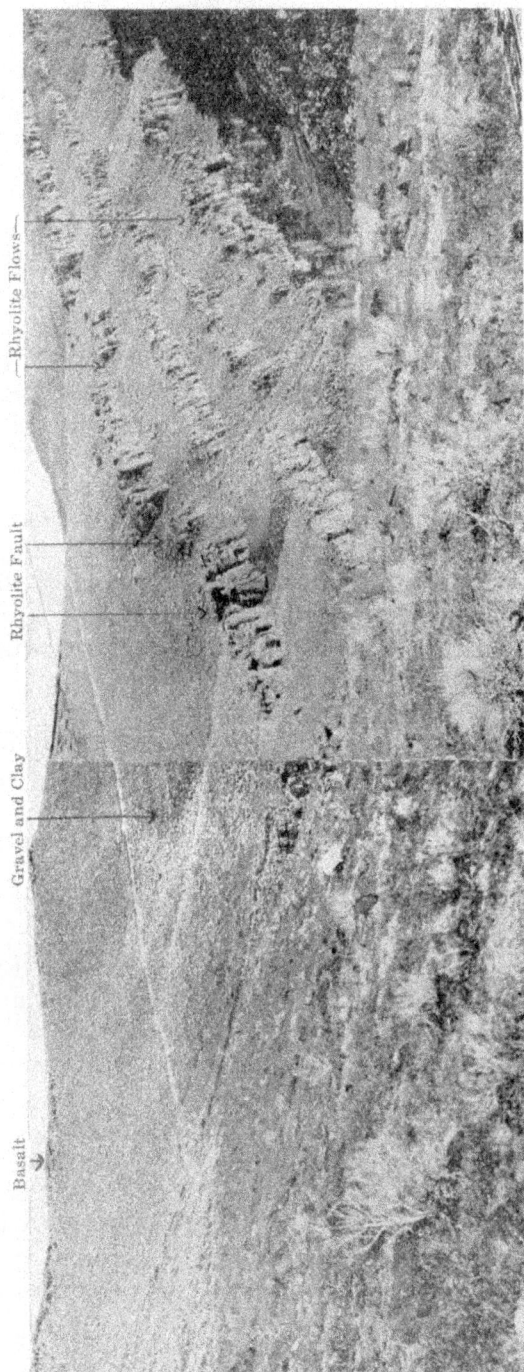

Figure 5. Tertiary gravel and clay beds north of Jarbidge, Nevada.

gold have never been found. The converse of this statement is
also true to some extent, and, likewise, many important lode
deposits, wherein gold is the principal metal have never produced any gold from associated placers. Considering the large
number of gold camps within the State it is remarkable how few
have had associated placers. Such well-known camps as Cherry
Creek, Seven Troughs, Bullfrog, Pine Grove, Delamar, National,
Ely (Lane City), Jarbidge, Goldfield, Divide, Silver Peak, and
others of lesser importance have produced little or no placer gold.
The production of precious metals in Nevada has been approximately one billion dollars, of which about 40 percent has been
in gold and the balance in silver. In comparison with the gold
production from lode mines the placer production of $30,000,000
appears very small, amounting to about 7 percent, particularly
in comparison with the State of California, where placer production has accounted for a much larger proportion.

Due to the fact that outcrops are situated at the surface where
changes are most rapid, they have had a more eventful geological
history than other portions of deposits. As the outcrop is eroded
away the component minerals are broken up and migrate either
mechanically or in solution. If they are moved mechanically, the
heavier portions may be concentrated as placers when conditions
are favorable. If conditions are unfavorable, they may be distributed widely and lost, as far as their economic value is concerned. If the minerals migrate in solution, they may enrich the
deposit lower down.

Many of the precious metal deposits in Nevada carry silver
as the principal metal; such deposits seldom have associated
placers as silver is chemically less stable than gold. Such important silver camps within the State as Tonopah, Fairview, Wonder,
Yellow Pine, Cornucopia, Spruce Mountain, Cortez, Eureka, Mineral Hill, Reese River, Candelaria, Ward, Belmont, Hamilton, and
a number of others, have never had any placer production. The
Comstock and Rochester Districts are exceptions, however. Placer
gold is nearly always alloyed with some silver. The insolubility
of placer gold as compared with silver has been shown by McCon-
nell[8] in nuggets from the Klondyke, which actually have a greater
fineness on the outside than the inside. The loss of silver on the
outer portions of nuggets was found to be from 5 to 7 percent
as compared to the inner parts.

Emmons has determined the relation of manganese to the

[8]McConnell, R. G., Report on the Gold Values in the Klondyke High-Level
Gravels: Geol. Survey of Canada, Ottawa, 1907, p. 979.

secondary enrichment of gold deposits. Finely divided gold
and the presence of manganese oxide are unfavorable to the
formation of placers and are favorable to leaching of the upper
parts of the oxidized zone. According to Emmons:[9] "Inasmuch
as enrichment is produced by the downward migration of gold
instead of by its superficial removal and accumulation, it should
follow that both gold placers and outcrops rich in gold would
generally be found in connection with nonmanganiferous deposits;
and this inference is confirmed by field observations. Placer
deposits are in general associated with nonmanganiferous lodes,
and such lodes are generally richer at the outcrops and in the
oxidized zones than in depth, the enrichment being due, in the
main, to the material associated with the gold."

In Nevada, with the exception of the Tertiary river channels,
it may be stated as a general rule that placer gold originated
from nearby lode deposits. The writer is of the opinion that
the gold in the placer deposits in the State that are of economic
importance has traveled, at the most, not more than 10 miles,
this distance being about the maximum length of the canyons
from their heads to the points where they debouch into the valleys. No doubt a large amount of fine gold has been transported long distances and is scattered in the sediments that fill
the valleys, but, with the exception of the gold in alluvial fans
at the mouths of some of the canyons, the gold in the valley sediments is too scattered to be of economic importance. In most
of the placer deposits within the State the gold has traveled less
than five miles.

Rounded or flattened particles of gold in well - washed and
rounded gravels generally have traveled farther from their
source than coarse rough gold with sharp edges and show more
or less crystalline structure. It may be stated that the angu-
larity of placer gold is roughly proportional to the distance it
has traveled from its source.

PROSPECTING AND SAMPLING PLACERS

The greatest incentive the placer miner has in searching for
gold is the possibility of discovering rich pockets in unsuspected
places. In searching for placer deposits in new areas, it is important that the prospector bear in mind that the deposits may have
little relation to the present topography, as the conditions under
which some Nevada placers were formed were totally different

---

[9]Emmons, M. H., The Enrichment of Ore Deposits, U. S. Geol. Survey Bull.
625, 1917, p. 316.

from what they are now. It may be taken for granted that scant returns will be had from reworking ground that was mined by the pioneers, for, as a rule, these miners were very efficient with the simple equipment they used. As the simpler methods have undergone little change, it is unwise to rework old placer tailings unless new and better methods and equipment of established efficiency are available. Many of the placer deposits of the State that were first worked by Americans were afterwards worked over by Chinese, who were content with smaller returns.

The miner's pan is the most useful of all the prospector's equipment in making a preliminary examination of a placer prospect. In such prospecting it is customary to pan the samples taken from gullies or small stream courses. Because the gold tends to concentrate on bedrock, the occasional use of a test pit or shaft is necessary in order to expose the stratum directly above bedrock. If such preliminary examination is encouraging, more systematic prospecting, either by sinking shafts or by drilling, generally follows at points where it has been indicated the most gold will be found. One should not be deceived by occasional rich samples, as it is possible to obtain pannings from selected parts of most deposits in this State that will show more than average amounts of gold. The misleading effect of high pannings must be discounted in determining the average value of the gravel. Although pay streaks may have a very high gold content per cubic yard, it is reasonably certain that this high value will not be maintained throughout the whole auriferous volume.

In small-scale placer mining operations it is not so important to sample thoroughly prior to installing equipment, because in most cases the richer gravel is worked as rapidly as it is uncovered. However, where it seems expedient to work the deposit on a large scale in order to make a profit, the average value per cubic yard of pay streak is important. It is safe to say that more placer mining ventures have failed for lack of thorough sampling than for any other reason.

Placer deposits, especially dry ones, usually may be investigated and sampled cheaply. The gold content can be estimated accurately by means of open cuts, pits, or drilling. If such an investigation be conducted properly it should disclose, with reasonable accuracy, the quantity of pay gravel available, the probable percentage of gold recovery, and indicate the type of equipment best suited to the work.

An unusual method of sinking holes has been employed on a

placer deposit near Atolia, Randsburg District, California.[10] A machine called a "dry rotary drilling rig" was used. It works on the principle of the ordinary post-hole augur, and was designed for digging wells and cesspools in the vicinity of Los Angeles. Similar equipment may be used advantageously to test placer deposits in Nevada, where conditions for its use are suitable. The machine used at Atolia is shown in figure 6.

The boring apparatus is mounted on the rear of a 3-ton automobile truck so that it is readily portable. It drills a hole about

Figure 6. Dry rotary drilling rig used for testing placer gravels, Atolia, California. (Courtesy E. & M. J.)

30 inches in diameter. The drill or Kelly rod is two inches square and is rotated by a ring gear, in which the rod is centered by a heavy yoke. The cylindrical bucket at the lower end of the Kelly rod is equipped with two curved cutting blades that are detachable for sharpening or renewal. The bucket revolves with the blades, and behind each blade is an opening through which the gravel is sliced into the bucket as the bucket revolves. When in operation, the bucket assembly revolves at a speed of 30 revolutions per minute. The weight of the drill assembly forces the

[10]Draper, Marshall D., Placer Ground Sampled by Well-Digging Equipment: Eng. and Min. Journal, vol. 133, 1932, p. 537.

blades to cut into the gravel, and in hard digging extra weights can be fastened to the rods.   Power for rotating the bucket is provided by the automobile engine through a clutch arrangement.

The drill and bucket are raised and lowered by a small winch and cable, the latter running through a sheave mounted on top of a digging ladder about 22 feet high.   The digging ladder is hinged so that it can be dropped on the cab of the truck when the latter is moved.   When the bucket is full it is raised clear of the hole and swung to one side.   A latch releases the hinged bottom of the bucket and permits the contents to be discharged.

The drill will readily cut unconsolidated gravel, providing the boulders are no larger than the width of the openings behind the blades, which is about six inches.   If a boulder larger than six inches is encountered it must be removed by hand, and for this purpose a man is lowered into the hole.   Sometimes it is easier or safer to start a new hole.

A dry rotary drilling rig was used in 1934 in digging government wells in the eastern part of Nevada.   This work was done under the drought relief program and some 50 wells were drilled for stock - watering purposes.   The deepest hole drilled was 110 feet.   This work was done at a considerable saving over the cost of hand operations.

With the dry rotary drilling rig, holes can be drilled in fine material under water by equipping the openings in the bottom of the bucket with flap valves.   Two men are required to operate a single drill, although a third man usually is necessary to assist in sampling, to sharpen the blades, and to do miscellaneous work.

The drilling at Atolia was done under contract by a Los Angeles company engaged in digging cesspools and sewers.   The contract specified that 1,000 feet of holes averaging 30 feet deep were to be dug at a rate of 35 cents per foot of depth.   The company expected to drill about 200 feet per day with two rigs, but, owing to the presence of some large boulders, the average progress was only about 150 feet per day.   The walls of the holes are smoother than those made by hand sinking and stand well without sloughing.

No arbitrary statement can be made as to the volume of gravel of known gold content that is necessary to insure the commercial success of placer mining.   Obviously, the richer the gravel, the smaller the tonnage necessary to net a profit.   Occasionally, large deposits of very low - grade gravel may be worked very cheaply to yield a good profit.

The gold content indicated by fire assays should not be employed in calculating the recoverable gold in placer ground, because gold enclosed in particles of quartz which would be included in the assay values would not be recoverable with the usual placer equipment. The gold content as determined by panning, or the use of the rocker, is more nearly comparable to results actually obtained in placer mining.

Table 2 affords a simple method by which a prospector may determine the approximate value of mine run of auriferous gravel per cubic yard. In using the table, a sample of one cubic foot of loose gravel is carefully washed in a pan or rocker and the weight of the recovered gold determined. If the gold is weighed in milligrams, this weight can be converted to grains by multiplying the number of milligrams by the factor .015432. The fineness of the gold usually can be estimated with sufficient accuracy from previous operations in the same area. In the table opposite the weight of the gold in grains and under the column of fineness will be found the value of the gravel per cubic yard.

Example: A sample of one cubic foot of loose gravel yields gold weighing 1.3 grains; the gold is 820 fine. The value per cubic yard for one grain of this fineness is $1.114 and for .3 grain $0.343, or a total of $1.457 per cubic yard in place. This value is based on $20.6718 per fine ounce for gold, and with the present price of $35 per ounce for gold the values per cubic yard as determined from the table must be multiplied by the factor 1.693, which gives a value of $2.47 per cubic yard.

TABLE 2

Prospectors' Table for Determining the Approximate Value per Cubic Yard of Auriferous Gravel (Based on $20.6718 per ounce for 1,000 fine)

Weight of gold from 1 cu. ft. of loose gravel (Grains)	VALUE PER CUBIC YARD IN PLACE FINENESS					
	780	*820*	*860*	*900*	*940*	*980*
.1	$0.109	$0.114	$0.120	$0.126	$0.131	$0.137
.2	.217	.229	.240	.251	.262	.273
.3	.327	.343	.360	.377	.394	.410
.4	.435	.485	.480	.502	.525	.547
.5	.544	.572	.600	.628	.656	.684
.6	.653	.687	.720	.753	.787	.820
.7	.762	.801	.840	.879	.918	.957
.8	.871	.915	.960	1.005	1.049	1.094
.9	.978	1.030	1.080	1.130	1.181	1.231
1.0	1.088	1.142	1.200	1.256	1.312	1.367
2.0	2.177	2.288	2.400	2.512	2.624	· 2.735
3.0	3.265	3.433	3.600	3.767	3.935	4.102
4.0	4.354	4.577	4.800	5.023	5.247	5.470
5.0	5.442	5.721	6.000	6.279	6.559	6.837

NOTE—In the above table the "swell" of the gravel is taken as 20 percent; namely, that 27 cubic feet of gravel in place is equal to 32.4 cubic feet when loose.

2

The pure gold content (fineness) of placer gold varies greatly in different districts and often in different parts of the same district. The fineness usually increases with distance from the original source of the gold. Placer gold is generally of greater average purity than that in the veins from which it was derived.

Table 3 gives the value per troy ounce of gold of different degrees of fineness.

TABLE 3

Value of Gold of Different Degrees of Fineness per Troy Ounce, Based on $20.6718 per Ounce for 1,000 Fine

Fineness	Value	Fineness	Value	Fineness	Value
600	$12.4031	740	$15.2972	880	$12.1912
610	12.6098	750	15.5039	890	18.3979
620	12.8165	760	15.7106	900	18.6046
630	13.0233	770	15.9173	910	18.8114
640	13.2300	780	16.1240	920	19.0181
650	13.4367	790	16.3307	930	19.2248
660	13.6434	800	16.5375	940	19.4315
670	13.8501	810	16.7472	950	19.6382
680	14.0568	820	16.9509	960	19.8450
690	14.2636	830	17.1536	970	20.0517
700	14.4703	840	17.3643	980	20.2584
710	14.6770	850	17.5711	990	20.4651
720	14.8837	860	17.7778	1,000	20.6718
730	15.0904	870	17.9845		

The values in the foregoing table are based on a rate of $20.6718 per fine ounce; at the current price of $35, the values given in the table must be multiplied by 1.693.

The average fineness of the placer gold mined in Nevada from 1906 to 1934, assuming that all the silver produced from the placers was alloyed with the gold, is 713.

PLACER MINING METHODS AND EQUIPMENT
INTRODUCTION

Paradoxical as it may seem, gold in placer deposits is the easiest and yet the most difficult mineral to recover. When gold is coarse its recovery by gravity methods is comparatively simple; but, on the other hand, it may be so finely divided that several thousand colors may be required to equal the value of one cent. Along the course of the Snake River, in Idaho, countless "process machines" have been tested in attempts to extract the flour gold from the gravels, but in general the attempts have been unsuccessful.

Placer mining on a scale exceeding hand methods is as much of a business venture as are other enterprises. The factors that govern the success of a placer mining enterprise can, to a great extent, be predetermined. Many failures in placer mining may be attributed to too much optimism and not enough business and engineering ability.

Placer mining is affected by geological conditions and geographical environment; climate, meteorology, topography, vegetation, and the ratio of volume of gravel to its gold content all have a bearing on the design, construction, and operation of placer mining equipment. The scarcity of water in Nevada, more than any other factor, is the most serious deterrent to large-scale placer mining, either by dredging or hydraulicking. The scarcity of water makes necessary a diversity of methods in working the alluvial deposits in the State, even when worked on a small scale. These methods include using machines that require a minimum of water, which is piped or flumed from a distance and then reused, or employing diverse pneumatic and centrifugal appliances. Very often processes and machines used in placer mining are designed by persons unfamiliar with the practical conditions under which the machines must operate, and, in consequence, many such processes and machines never get beyond the "idea" stage. The number of gold-saving machines invented runs into the thousands, but most of them have been discarded and only a few that have demonstrated their worth have been adopted as being valuable in placer mining.[11] The best way to test the utility of a novel apparatus is to work it under field conditions on a scale sufficiently large and for a period long enough to demonstrate fairly what the machine can do.

There are a number of placer deposits in Nevada that have been skimmed of their richer gravels by hand methods, and the residue is of too low grade to be worked by hand and too remote from water sufficient for hydraulicking or sluicing. In such instances the problem is to find a means for treating large yardage at low cost. The equipment for handling the gravel must be very efficient, must require little water, or must be able to reclaim used water. Drag lines, excavators, tractors, and scrapers or small revolving power shovels are coming into use, and the application of such machinery to placer mining makes for cheap mining. Equipment should be selected to fit local conditions which has not always been done. No matter how successful the metallurgical equipment may be, the machinery used to handle the gravel must operate profitably at low cost. The real basis upon which to choose placer-mining machinery is its capacity to work with a high average of efficiency.

The impression exists that placer mining requires a comparatively small outlay of capital. This is true for small-scale placer

---

[11] Descriptions of over 150 processes and machines for recovering placer gold are given in the April and July, 1934, issue of the California Journal of Mines and Geology, published by the California Division of Mines, Ferry Building, San Francisco.

mining, but when the amount of material handled is large, expensive equipment is necessary. Placer mining consists essentially in excavating, transporting, screening and washing sand and gravel and disposing of tailings. When this is done on a large scale expensive equipment is required. Under favorable conditions, and with dredging or hydraulicking operations carried out on a large scale, gravel containing less than five cents in gold per cubic yard can be worked at a profit, but this is not true in Nevada.

Writers on placer mining have stated that the three essentials for successful placer mining are sufficient gold in the gravel, water in abundance, and plenty of dump room. Water is seldom present in abundance at Nevada placer deposits, yet numerous placer operations have been carried on successfully in this State.

Much gold has been taken from arid regions in Mexico and Australia and from the dry placers of Nevada, California, Arizona, and New Mexico by means of dry placer mining. To one who is familiar with the operation and efficiency of the Mexican air jig, or "dry washer," it is most surprising to receive such misinformation as the following, which is contained in a treatise on hydraulic and placer mining, that has run through several editions: " * * * it may be set down as an axiom that all dry placer machines will prove failures, unless the gold is so plentiful it may be sifted from the dirt, and under the latter condition a 10-mesh sieve would suffice." This is not true. Dry placer machines are capable of treating material from which gold cannot be recovered by screening. However, the writer on hydraulic and placer mining qualifies his conclusions when alluding to a patented pneumatic placer machine: "The writer not being particularly interested in that kind of mining, has never inquired into its virtues or where it has been successfully used." The ability of the skillful dry-wash operator to save gold has been proved countless times. An example is cited in the record of Thomas ("Dry Wash") Wilson, known throughout Nevada. Wilson won a fortune from the placer gravels at Round Mountain in 1907 and is an expert in dry placer-mining practice.

In the following pages various types of small - scale placer-mining equipment that have been found suitable in Nevada or in other fields are described and their limitations are discussed. Additional data on placer mining equipment are included in the chapters on the various districts.

THE PAN AND BATEA

The gold pan is a flat-bottomed, circular dish made of iron, copper, aluminum, or agateware. Some have rims of smooth steel and bottoms of copper, which may be silver - plated for amalgamating. The pan most commonly used in Nevada has a top diameter of 16 inches and a depth of 2½ inches, with sides sloping about 40 degrees. A riffle, formed by crimping the side inward about halfway up for a distance extending halfway around the circumference, is a useful feature.

The batea is similar to the pan, but is made of a single piece of wood and instead of having a flat bottom is either conical or rounded. The latter type is much like an old-fashioned kitchen chopping bowl, while in the former shape the angle of the conical bottom is about 150 degrees. Mahogany is a good wood for bateas and should be turned in a lathe with the direction of the grain normal to the surface. The batea is especially useful for saving fine gold, which tends to adhere to the wood surface but slides over metal. In Nevada the batea is used less frequently than the pan.

Due to its small capacity, the pan is used in Nevada mainly in sampling, prospecting, and cleaning concentrates. Placer ground must be very rich in order to make panning alone profitable, for an experienced panner can only treat about one cubic yard per day.

THE ROCKER

Rockers are in general use in the State and they vary considerably in size and form. Some are unnecessarily elaborate, with beveled joints and flaring sides or ends, and are difficult to construct. As a rule, the simpler designs are as efficient as the more elaborate ones. All rockers are constructed on the same basic lines and consist essentially of a riffled bottom, a movable screen for discarding the larger material, and a deflecting apron to direct the flow of gravel to the rear of the rocker and at the same time assist in catching the gold. In most cases the rockers are manipulated by hand, although some are equipped with power. The rocking motion is necessary to concentrate the gold in the riffles and keep the gravel moving.

Some rockers are constructed with a single apron, while others have two. In addition to riffles made of wood strips, various materials are used in rockers as gold catchers, including carpet, canvas, rubber mat, and amalgamated copper plates.

An excellent and simple type of rocker is described by H. H. Symons of the California State Division of Mines in the 1932 report of the California State Mineralogist (figures 7 and 8). The dimension of parts are tabulated as follows:

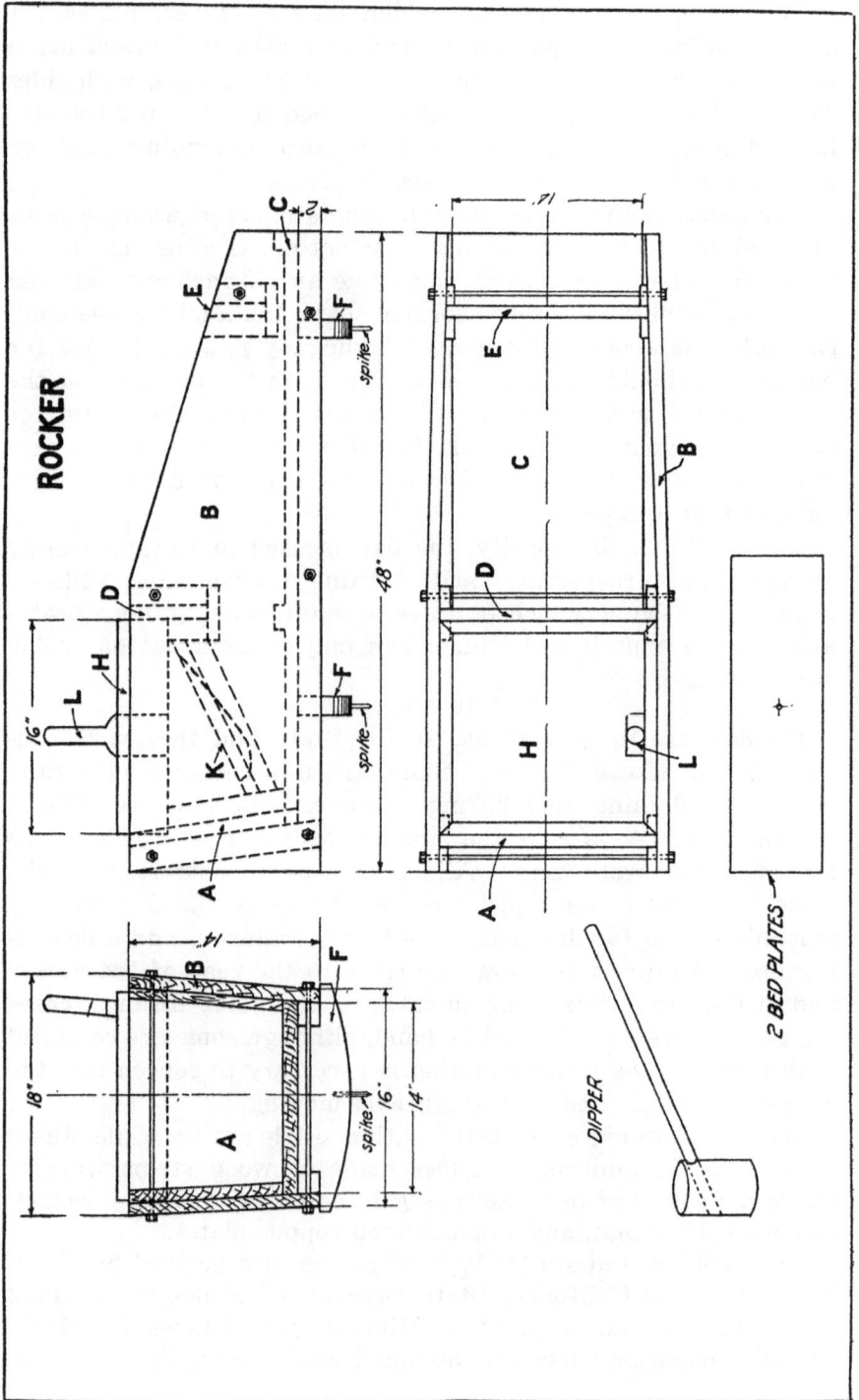

Figure 7. Rocker assembled.

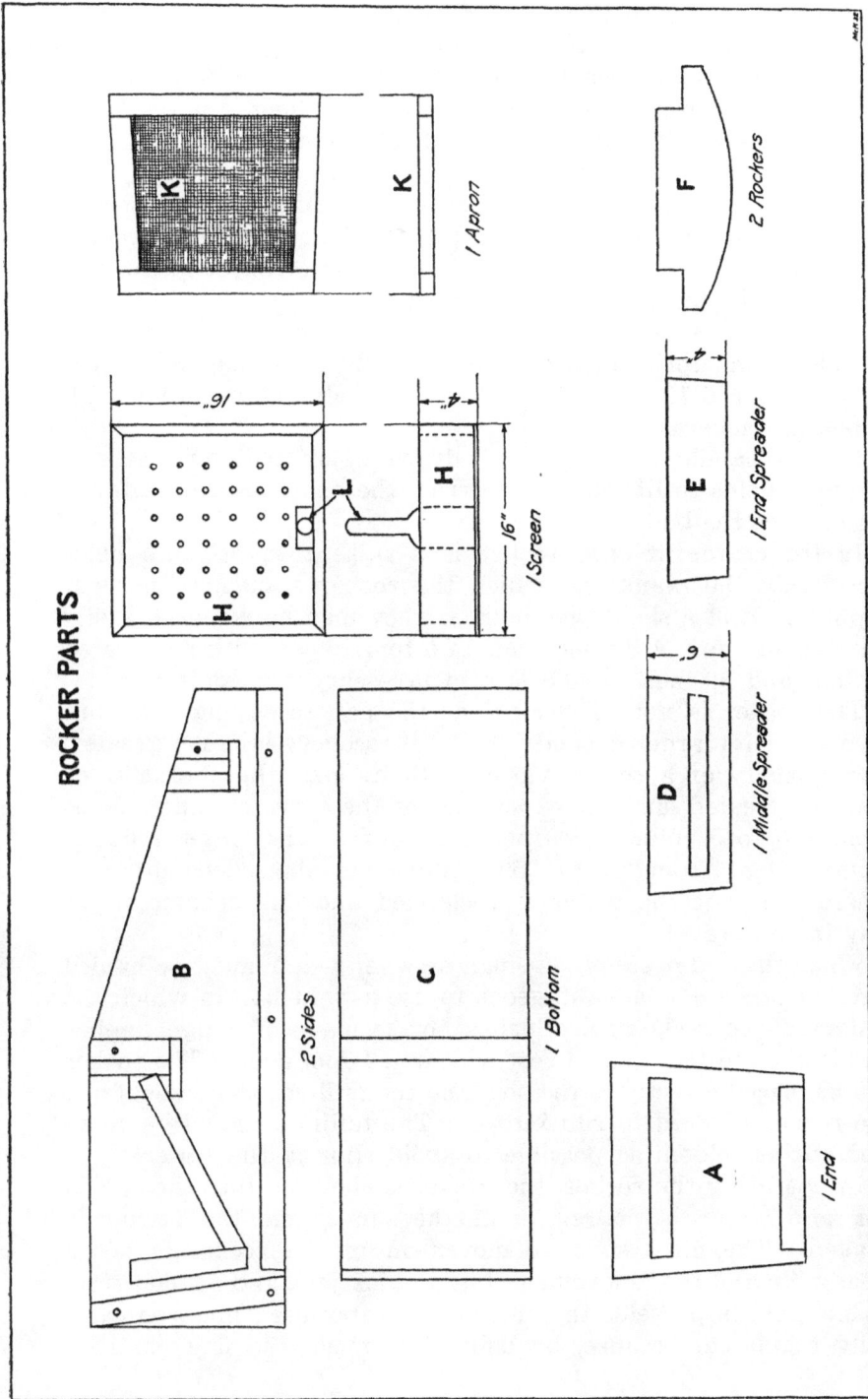

ROCKER PARTS

K — 1 Apron

F — 2 Rockers

H — 16" — 1 Screen

E — 1 End Spreader

D — 1 Middle Spreader

B — 2 Sides

C — 1 Bottom

A — 1 End

Figure 8. Parts of rocker shown in Figure 7.

A—End, one piece 1″ x 14″, 16″ long.
B—Sides, two pieces 1″ x 14″, 48″ long.
C—Bottom, one piece 1″ x 14″, 44″ long.
D—Middle spreader, one piece 1″ x 6″, 16″ long.
E—End spreader, one piece 1″ x 4″, 15″ long.
F—Rockers, two pieces 2″ x 5″, 17″ long.
H—Screen, about 16″ square outside dimensions with screen bottom. Four pieces of 1″ x 4″, 15½″ long and one piece of screen 16″ square with ¼″ or ½″ openings or sheet metal perforated with similar size round openings.
K—Apron, made of 1″ x 2″ stubs covered loosely with canvas. For cleats and apron, etc., 27 feet of 1″ x 2″ is needed. Six pieces of ⅜″ iron rod 19″ long threaded 2″ on each end and fitted with nuts and washers.
L—The handle, in Mr. Symon's drawing, is fitted on the screen, where it helps to lift the screen from the body. Some prefer it fastened to the body.

In the center of each rocker is a spike that fits into holes bored into the planks on which the rocker stands, to prevent slipping. Riffles should be made so they may be removed easily for cleaning up. A dipper made of a tomato can with perforated bottom and 30-inch handle is also necessary for adding water.

The rocker is not as efficient as the pan in saving gold, but because of its larger capacity the total recovery is much greater. The capacity of a rocker varies with its size, the character of material treated, and the experience of the operator, and ranges from 2½ to 5 cubic yards per man shift. The water requirements range from 75 to 250 gallons per day, depending on whether or not the water is reclaimed and on the amount of clay in the gravel.

When the water supply is small or when water must be hauled some distance it is advantageous to use a tight box in which the water can be reclaimed. Such a box is made of rough lumber and is about 6 feet long, 4 feet wide and 1 foot deep. The rocker can be placed directly in the box, and the tailings and water from the rocker allowed to run into it. The tailings should be shoveled out as seldom as possible to avoid sliming the water.

In operating the rocker, the gravel is shoveled into the screen box and water is poured in at the same time the rocker is shaken. The most effective movement of the rocker is not a steady "to-and-fro" movement, but a quick jerk and sudden stop. If fine gold is present, the amount of water used must be carefully regulated, because, by using too much, the fine particles of gold may be washed over the riffles. Black sand should not be allowed to accumulate to the top of the riffles but should be removed. The apron should be washed off into a tub at the same time. To assist in recovering fine gold, the lower part of the rocker may consist of an amalgamation plate on which mercury has been rubbed, and the canvas apron may be replaced by a piece of blanket or other material with a nap.

Where the auriferous gravel is cemented loosely or contains considerable clay, it is advisable to use a puddling tub in order to disintegrate the material thoroughly before treating it in the rocker. This practice saves time and makes the working of the rocker much easier. A puddling tub can be made of the bottom half of a large wooden barrel. The gravel is disintegrated by mixing it with water in the tub and stirring it with a rake. A plug several inches above the bottom is convenient for drawing off the clay held in suspension before the gravel is shoveled into the rocker.

THE LONG TOM

There are several forms of long tom, the usual one being an inclined trough or modified sluice box. The long tom requires more water than the rocker and, although of larger capacity, it is not so efficient.

A sketch of a long tom is shown in figure 9. Ordinarily the long tom is built in three sections. The upper section is a sluice box 6 x 12 feet long, 8 to 10 inches deep, and 12 inches wide, the bottom of which is lined with sheet iron. The middle section is 6 to 12 feet long, about 20 inches wide at the upper end and increasing to 30 inches at the lower. This section also is lined on the bottom with sheet iron. The bottom of the widest portion consists of a strong, perforated sheet iron screen or tom iron, which is inclined at an angle of 45 degrees. The lower section, or riffle box, is 6 to 12 feet long by 36 inches wide. Riffles and canvas are placed in the bottom of this section to catch the gold. The grade of all three sections is about 1 inch per foot.

The long tom is placed in position as near the ground as possible and the water allowed to enter at the upper end, either through a hose or trough. Two or more men shovel the gravel into the upper end, and the water carries it downward to the middle section, where a third man stirs the gravel about on the tom iron until all except the larger stones have passed through. These are forked out and thrown aside. The smaller particles of gravel and gold pass through the tom iron and into the riffle box, where the gold is held by the riffles while the gravel

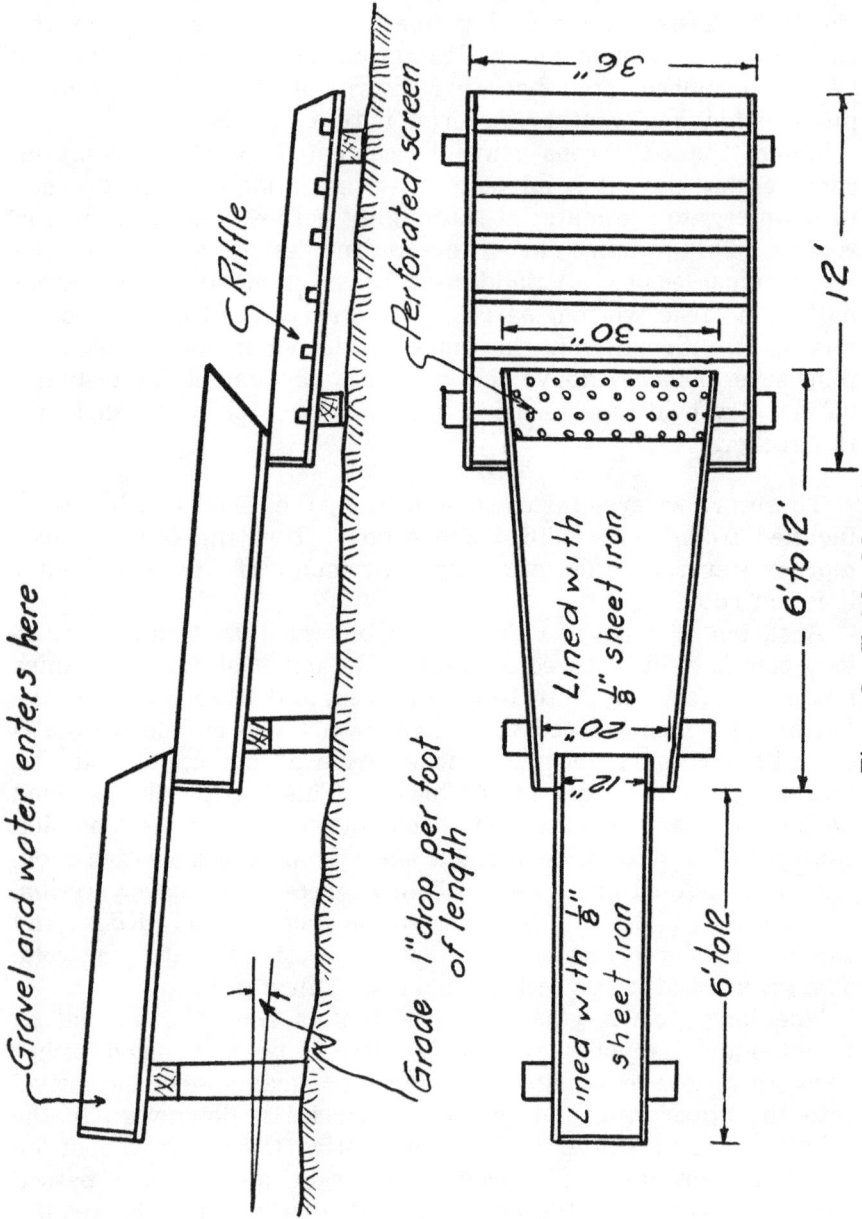

Figure 9. Sketch of long tom.

passes over the lower end. The long tom is not recommended when only a small amount of water is available. Owing to the amount of water required to operate it the long tom is seldom used in Nevada. A disadvantage of the long tom is that the riffle box is too short and some of the finer gold particles do not have time to settle.

THE SLUICE

The sluice is probably the most effective small device for separating gold from gravel with the least amount of energy. The sluice is used widely in Nevada, particularly during the spring season when more water is available than at any other time of the year. The sluice generally consists of a wooden launder equipped with riffles, through which the gravel is transported by running water. A sluice may also be a ditch cut in bedrock, and is then named a "ground sluice." When constructed of wood, the size of the sluice depends on a number of factors, such as the character of the gold (fine or coarse) ; the physical properties of the gravel (fine, coarse, smooth, angular, or clayey) ; the quantity of gravel; the quantity of water available; and the gradient and dump room.

For hand shoveling operations, a wood sluice ordinarily is 1 foot wide, 8 to 12 inches deep, and several "boxes" in length. As used in the West, a sluice "box" is generally 12 feet long. The sluice boxes may be constructed as a continuous flume or in 12-foot lengths. In the latter case, one end of the box is made narrower than the other so that the narrow end of one may telescope into the wide end of the one below. This sectional construction makes handling easier when it is necessary to move a number of boxes to a different locality.

The grade of the sluice varies from four to eight inches per box. When carefully constructed, the swelling of the lumber when wet usually is sufficient to make the bottom of the sluice tight, so that no gold can escape.

The bottom of the sluice is covered with riffles, which may be constructed of various materials. Several types of riffles are made for the small sluice. Angle iron, plank with two-inch staggered auger holes, woven wire screen, transverse crossbars of either wood or metal, wooden blocks placed endwise, or small rails placed longitudinally in the bottom of the sluice are used alone or in combination. The relative advantage of each type of riffle is largely a matter of individual opinion; in most cases the selection of the type of riffle depends upon the material most readily available.

Riffles should not be fastened permanently in the sluice, as it is necessary to remove them for the clean-up. Transverse riffles may be constructed in sections and held in place either by their own weight, when made of iron, or by wedges, when of wood.

At the lower end of the sluice, the finer material may be screened out and run over a blanket, cocoa matting, or other material adaptable to the recovery of fine gold.

In operating the sluice, the gravel is shoveled into it and the larger boulders removed either by hand or with a fork. In some cases the oversize material is removed by shoveling the gravel onto a grizzly at the head of the sluice. In some operations

Figure 10.   Power-driven sluice box used at Lynn, Nevada.

mercury is placed behind the riffles to catch the fine gold. When cleaning up, the riffles are removed and a small stream of water is allowed to flow down the sluice to wash the heavy sand. The nuggets are picked up by hand and the black sand concentrate collected and washed in a gold pan or rocker. The black sands, if magnetic, may be separated from the gold by a magnet after the concentrate has been dried. To prevent fouling of the magnet, a piece of cellophane may be interposed between the magnet and the magnetic material.

A modified type of sluice, which has proved very satisfactory in Nevada, is simply a box 12 to 16 feet long, 12 by 12 inches in section, made of 2-inch plank, and mounted on springs or a shaft so that it can be rocked from side to side or endwise by a small gasoline engine. The writer prefers the side-shake type, as in it any cemented material may be broken up better. The stroke of the machine is about three inches. A machine of this type is shown in figure 10. The bottom of the sluice box may be equipped with transverse or longitudinal riffles; sometimes woven wire screen is used. The water and gravel are fed into the upper end of the box and the tailings are discharged at the lower end. The box is set nearly level and, if the gravel contains clayey material, boulders or steel balls may be placed in the box to aid in breaking up the lumps. The capacity of a machine of this type varies from 1½ to 2 cubic yards per hour.

EXTRACTING GOLD FROM AMALGAM

If mercury is used to catch gold in placer mining operations the resulting amalgam may be retorted in order to separate the gold from the mercury. Retorts can be purchased from dealers in chemical supplies. When the amount of amalgam is small, the quicksilver may be driven off simply by heating the amalgam on a shovel or iron receptacle placed over an open fire or forge. The mercury is volatilized and lost. Amalgam should be reduced with great care as mercury fumes when inhaled are very poisonous.

A simple expedient used by prospectors to extract the gold from amalgam is to place the amalgam on a shovel or iron plate and cover it with half of a raw potato, in which a hole has been scooped out. The shovel or plate is placed over a fire and, after heating it for about 20 minutes, the gold sponge is found on the iron. Some of the mercury can be recovered by breaking up the potato and panning the pulp.

DRY WASHERS

The recovery of gold from placer gravel by dry concentration is a subject that has received much attention in the arid regions where water is too expensive to be used in orthodox placer-mining methods. Inventive minds have been occupied with this subject for years, and a great many machines have been designed for this work. It is interesting to note that the late Thomas A. Edison, about 1897, designed a dry process machine for saving gold from a placer deposit in New Mexico.[12] Most of the machines, however, have been short-lived, although some have proved their worth under certain conditions.

In Mexico, placer gravel has been concentrated by the crude

---

[12]Chapman, Cloyd M., The Edison Dry Process for the Separation of Gold from Gravel: Eng. & Min. Jour., vol. 75, 1903, p. 713.

method of exposing it to the action of the wind while rolling and tossing it in a blanket. The finer particles are blown away and the coarser material is picked over by hand, while the fine gold is enmeshed in the hair of the blanket. A method of dry panning combined with hand picking, winnowing, and blowing with the breath has been employed in working rich placer areas in the desert regions of Australia.

Machines designed for the concentrating minerals, using air as a medium, may be divided into three groups, depending on how the air is used, as follows:

1. Machines that project the material to be concentrated through air by force other than an air blast, the particles being classified according to their momentum.

2. Machines using a continuous blast of air.

3. Machines using intermittent pulsations of air, whereby the heavier mineral particles settle to the bottom and the lighter material comes to the top and is removed by gravity and air pulsations.

Machines of the first type are little used. Among the dry concentrators that employ a continuous blast of air may be mentioned the Cottrell, Stebbins, and the Sutton, Steele and Steele, tables. The so-called "Mexican air jig" is perhaps the best known example of a pneumatic machine using intermittent pulsations of air for concentration.

There is nothing fundamentally unreasonable in expecting good results from dry concentration in limited application; however, there has to be some justification for resorting to dry treatment. Whichever method of dry concentration is employed in treating placer-gold material, the operator is confronted with the drawback that the difference in density between gold and gravel is less in air than in water. This fact works against the use of any machine employing a medium lighter than water for concentration. As an illustration, the specific gravity of gold in air is 19.3 and that of quartz 2.65, the ratio being 19.3/2.65, or 7.28; when submerged in water, the density ratio is (19.3–1)/ (2.65–1), or 11.09. In recovering fine gold from placer gravel other factors intrude and the problem becomes increasingly difficult. In general, it is easier to effect a separation of two minerals of different specific gravities when the operation is carried on with water, as there is a greater separating effect between the heavy and light components of the material.

In addition, dry concentration requires that the material be perfectly dry before concentration can be effected. The physical character of the gold also is important because, if the gold is in the form of "flour," it cannot be caught with the usual placer equipment. Water disintegrates clayey material, freeing the particles of gold for treatment while such material is affected little by air. When using water for concentration, the products, graded according to size, are treated more easily; with air the concentrating machine must work effectively on material of all sizes ranging up to half an inch. Furthermore, concentration with air introduces a dust problem, which is troublesome; the fine dust that is stirred up is irritating to the eyes and injurious to the lungs.

Since air diffuses and water does not, air concentration lacks one of the principles that aids hydraulic jigs to be self-regulating. A hydraulic jig can have a sieve that permits grains of smaller diameter to be discharged into the hutch and the coarser material, which belongs in the tailings to be discharged as such. The air jig, on the other hand, cannot have a bed that will act in this manner and is handicapped by its lack of ability to use suction.

The lightness of air permits a higher number of pulsations per minute with it than with water. The principal advantage of air is its accessibility in dry climates where water usually is scarce.

Dry washing machines of various designs and sizes are used in working the placer deposits in Nevada, where the water available is insufficient for wet methods. Despite the large number of dry washing machines that appear from time to time, the type most generally used is the Mexican dry washer or air jig, which is especially satisfactory for small placer mining operations as it is simple to construct. Although there are a number of different styles of this machine, the same principle is employed in the construction of all of them. The dry washer operates on the same principle as the jig, and consists essentially of a screen, hopper, riffle board, and bellows, all mounted on a wooden framework so as to be easily portable. A drawing of such a dry washer, kindly furnished by Mr. Ben Hood of Reno, is shown in figure 11. The bellows at the bottom of the frame is made of canvas. Such a bellows has a life of about 150 cubic yards of gravel. The air enters the bellows at the bottom through several holes covered with leather flap valves. The top forms a stationary riffle board, which is removable. The riffle board is covered with ordinary window screen, on top of which is stretched either canvas, linen, or muslin. On top of

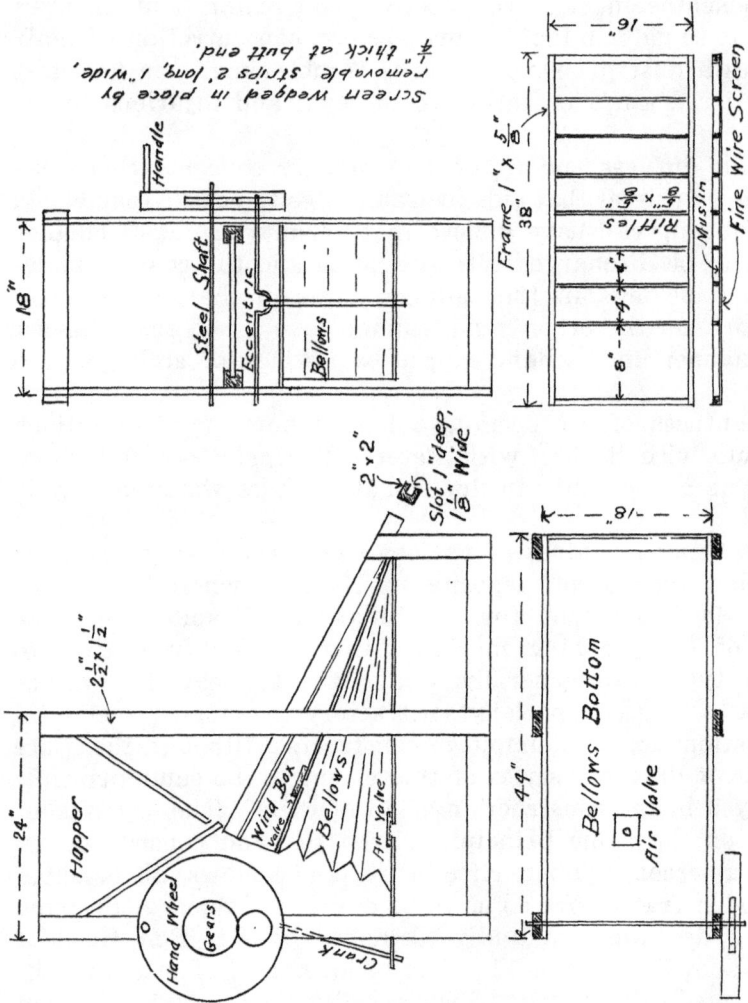

Sketch of Nevada Type Dry Washer

Figure 11. Sketch of Nevada type dry washer.

the cloth are transverse riffles spaced from four to six inches apart. The bellows is actuated by hand crank and eccentric to force intermittent gusts of air through the riffle board. Usually the gear ratio is such that one revolution of the crank works the bellows three times. Dry washers of this type are also operated by small power units, and several of these machines are described in the section of this paper dealing with the various districts.

Figure 12 shows another type of a home-made dry washer used in the Tenabo District. The cost of constructing a dry washer, including labor and materials, varies from $25 to $50. The capacity of a single machine depends on the character of the gravel, the physical conditions of the gravel and the manner in

Figure 12. Home-made dry washer used in Tenabo District, Nevada.

which the gravel is prepared for treatment. A hand-powered machine can handle from 1½ to 4 cubic yards per day.

The angle of inclination of the riffle box is fixed in all of the machines seen in operation by the writer, as experience has shown that the working angle varies but slightly. If it is found necessary to alter the angle of inclination, it may be done by slightly tilting the whole machine and fixing it in position with supports under the rear end. The slightly inclined riffle box will save more gold than the steeply inclined one, but less material can be treated in a given time in the former.

A design of a type of dry placer blower used in Australia is shown in figure 13. A is a shaking screen of punched sheet iron to which a reciprocal shaking movement may be imparted by hand. The undersize passes through to metal chutes, B, and the oversize is discharged over the edge. The chutes, B, deliver to a removable tray, C, provided with riffles and a bottom of coarse cloth or toweling backed by metal screen. This tray may be 10 or 12 inches long and 6 inches wide, or larger, and sits in the top of a wind box, D, having clack valves in the bottom, through which air from the 18- or 20-inch bellows, E, enters. The wind box sits on the bellows. The latter is operated from a screen A, through a system of levers F. Screen and bellows are set in a steel frame at suitable angles. In the type of blower

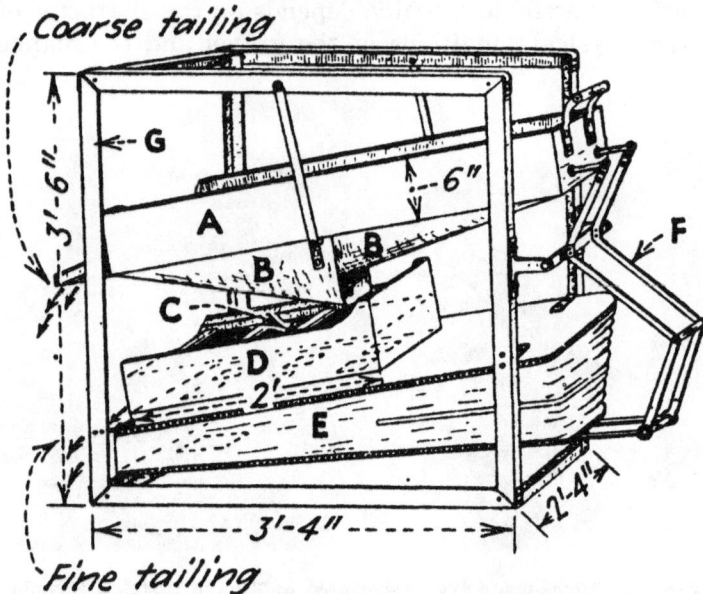

Figure 13. Type of dry blower common in Australia.
(Courtesy E. & M. J.)

shown in figure 13 the motion is longitudinal, but in other designs it may be transverse. When properly designed, this machine is said to be capable of recovering fine gold.

Still another type is the Hungarian dry washer, simple in design, belt driven either by hand or power, shown in figure 14. It has been in use in Europe more than a century. A machine of average size, worked by two men, is said to have a capacity of 15 to 20 tons per day. La Cienega, Sonora, Mexico, was the scene of intensive operations of the Serna family from 1884 to 1894, who employed 10,000 laborers to work the rich gravels with Hungarian dry washers.

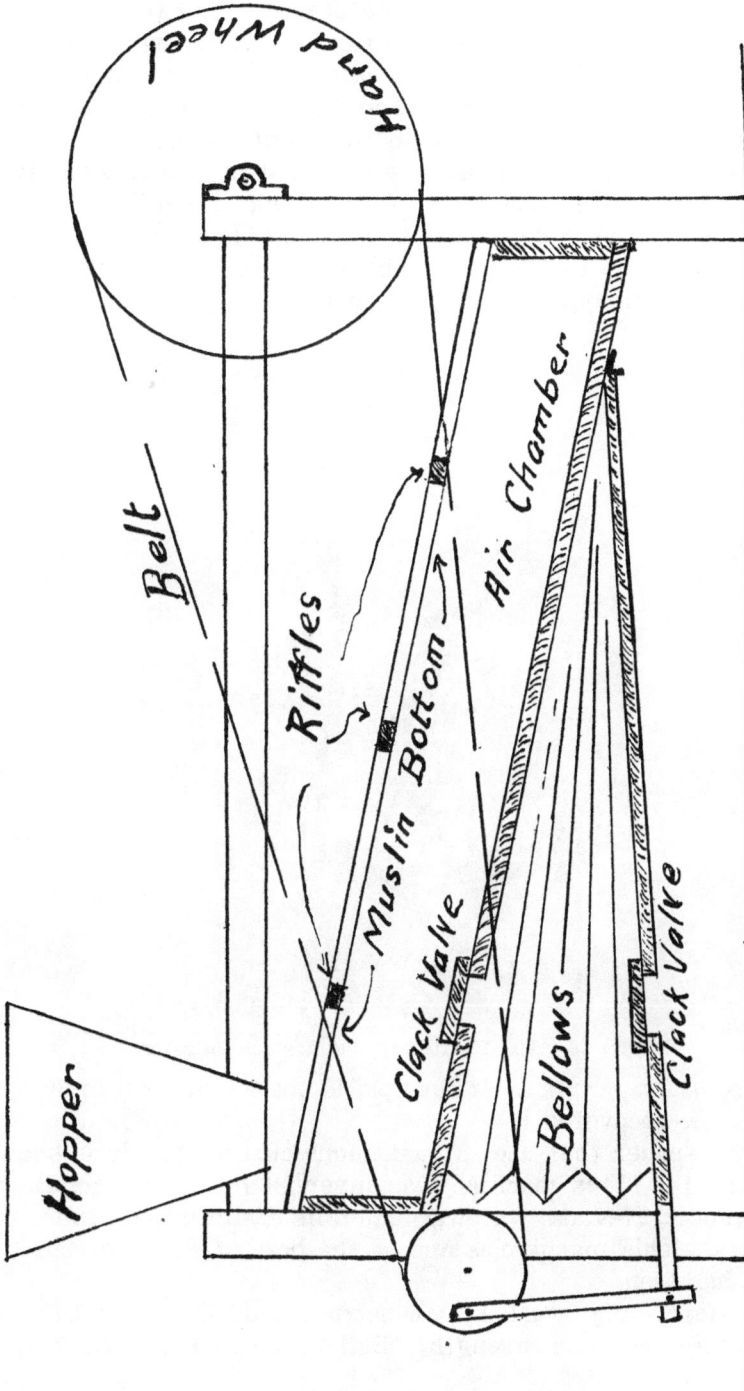

Figure 14. Sketch of Hungarian dry washer.

Either one or two men operate a machine, two being more satisfactory because one man can turn the crank while the other shovels the alluvium onto the screen.   The intermittent blasts blow the light particles of sand and dust away, while the coarser material works down the riffle board and discharges at the lower end.   The gold and black sands are caught by the riffles.   After two or three wheelbarrowfuls of gravel have been run through the machine, the riffle board is taken out and the concentrate dumped on a sheet of canvas.   This concentrate is cleaned either by rerunning through the machine when a sufficient quantity has

Figure 15.   The Harris portable dry concentrator.

collected or by panning.   If the gold is not too fine, up to 80 percent can be recovered.

A dry washer that uses a continuous current of air is shown in figure 15.   This machine was invented by W. H. Harris of Winnemucca, Nevada, who worked on its design at odd times for 10 years.   This machine is one of the best of its kind that the writer has seen.

The Harris dry washer is constructed of three-ply Douglas fir for lightness and strength.   Ball bearings are used, and it can be operated by hand or a half-horsepower gasoline engine. The weight of the machine shown in the accompanying figure is 45 pounds.

The auriferous gravel is shoveled or fed upon a quarter-inch mesh inclined screen placed on top of the machine. The over-size is discarded by gravity and the undersize passes into a hopper, from which it is fed automatically down upon the fixed riffle screen below. A continuous current of air produced by a wooden-paddle fan is forced through the riffle screen. The main feature of the machine is the baffle plates which are placed in the fan box below each riffle. The baffles deflect the air currents against the direction of flow of the gravel down the screen in such manner as to keep a space above each riffle clear of material to allow concentration of the heavier particles.

The riffle screen is 30-mesh and no cloth is used above it, as is done in the air jigs previously described. Gold finer than 30-mesh that may work through the riffle screen is recovered in a trap placed at the bottom of the fan box. The current of air through the screen is sufficiently strong to prevent much fine dust from dropping through the screen into the fan box.

The concentrates from the machine are cleaned by panning. Mr. Harris has also designed a similar machine with two fans and two riffle screens. The two riffle screens are placed one above the other so that any fine gold lost from the upper screen can be recovered on the lower one. The machines are equipped with dust deflectors so that the dust is carried away from the man who is turning the crank.

Over 100 of these machines are in use. In Sonora, Mexico, a battery of 12 machines has been hooked up to a single line shaft geared to another shaft turned by a mozo.

THE QUENNER DISINTEGRATOR

The Quenner disintegrator is a simple, inexpensive device for breaking up cemented gravel prior to sluicing or dry washing. It was used first in working the Altar dry placers in Sonora, Mexico, about 1910.

A sketch of the machine is shown in figure 16. The disintegrator consists of a trommel screen and a number of hammers mounted on a shaft independent of the trommel. A large machine, capable of handling from 250 to 300 cubic yards per day, is made up of a trommel 6 feet long, 3 feet in diameter, with holes 2 inches in diameter. The trommel is mounted horizontally on flanged wheels and is revolved by chain-and-sprocket drive at about 25

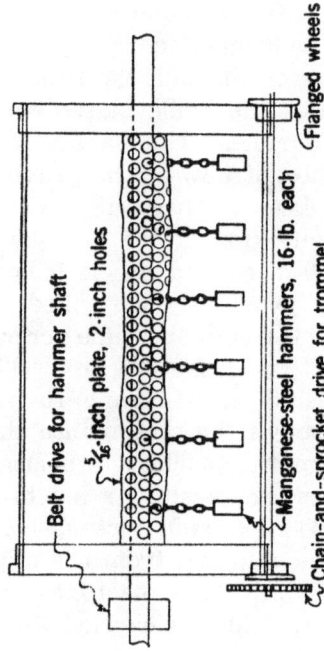

Figure 16. Quenner disintegrator.

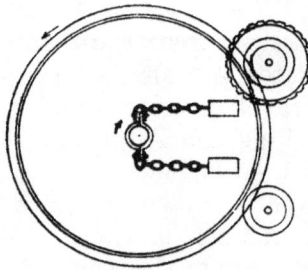

revolutions per minute. A belt-driven steel shaft $3\frac{7}{16}$ inches in diameter passes through the trommel and turns on its own journals. This shaft is belt driven. Attached to the shaft by bolts and clamps are 12 chains, on the ends of which are fastened manganese steel hammers weighing 16 pounds each. The shaft turns counter to the trommel at a speed of about 50 revolutions per minute. The revolving hammers and trommel disintegrate the gravel and carry it through the machine rapidly. Although the centrifugal force holds the chains to a semirigid position when running, they do not break when a boulder too hard to shatter is encountered, as would be the case with rigid rods.

The trommel undersize is diverted to the sluice box or other washing equipment, while the oversize is discharged to waste.

MECHANICAL WASHING PLANTS

Mechanical gravel washing plants other than dredges, used in Nevada, may be divided into two classes—portable and stationary. The equipment of both types is similar in design and consists essentially of trommel screen, sluice boxes or other gold-saving equipment, and a means for storing tailings. The motive power for the washing plant machinery may be Diesel engine, gasoline engine, or electric motor. The portable plant is mounted either on skids or wheels and is towed behind the excavating equipment.

The principal advantage of the portable plant is that it eliminates trucking charges. The portable plant, however, has a number of disadvantages that may outweigh any saving made in transportation costs. It may develop structural defects that are not present in the stationary plant owing to the fact that the size and weight of the plant must be kept to the minimum. The storage capacity, for example, is small so that not enough gravel can be stored to take care of delays caused by failure of the excavating equipment. Other drawbacks of the portable plant are the difficulties and delays encountered in leveling, making water connections, disposing of tailings, and moving ahead. Cleaning of bedrock also is hampered to some extent with a portable plant. The stationary plant, on the other hand, does not have these disadvantages and in addition there is more flexibility in operations.

A number of washing plants are described in this bulletin in the section on the various districts.

MECHANICAL EXCAVATING EQUIPMENT

In recent years mechanical excavating equipment has been used more and more in working placer deposits in Nevada. At present, operators have a wide selection of excavating equipment which the old-time miner did not have. The factors that govern the choice of equipment are both economic and technologic. Several failures in placer mining may be attributed to the fact that the excavating machinery was selected largely on the basis of first cost without due regard to the conditions under which it was to be used.

The various types of equipment that may be employed in excavating placer gravels are: Animal-drawn scrapers; tractor-drawn scrapers; power shovels; drag-line excavators; power scrapers; slackline cableways; hydraulic monitors; and dredges.

Accessory transportation equipment may be used in combination with some of the foregoing types, while other types operate in the dual capacity of excavating and transporting equipment. Equipment used solely for the transportation of material may include mine cars operating on track, automobile trucks, and conveyer belts.

The real basis upon which to select placer mining machinery is the capacity to continue working with a high average efficiency. Any portion of the plant that must remain idle increases operating costs, hence it is necessary to reduce the number of delays for any reason to a minimum.

The horse- or mule-drawn scraper combines excavation and transportation. Teams and scrapers are used in small - scale placer-mining operations where the deposit is shallow and the haul fairly short. The principal advantages of the animal-drawn scraper are low capital cost, simplicity and flexibility. Disadvantages are the relatively high operating cost due to limited capacity and the fact that material can be elevated only short distances, limited by the height of ramp construction. Because of their slow traveling speed, the range for animal-drawn scrapers is limited to several hundred feet. All scrapers are limited to easy cutting material unless it is plowed first. Animal-drawn scoop scrapers range in capacity from about 3 cubic feet for a single animal to 7 cubic feet for a team. The Fresno or Buck scraper, traveling on runners, ranges from 8 to 18 cubic feet capacity. The wheeled scraper which carries the load off the ground, ranges in capacity from 9 to 27 cubic feet. The actual capacity of a scraper is somewhat less than its measured capacity, because of the spillage.

The tractor-drawn scraper is a mechanized adaptation of the animal-drawn scraper. It is subject to virtually the same limitations. Greater speed is possible with the tractor-drawn scraper, hence its range of operation is somewhat greater. Owing to the fact that the power units are larger, the size of the scrapers can be increased. The original cost is higher than for the horse-drawn scraper, but the upkeep expense per unit of material handled is probably less under average conditions.

The power shovel operates to best advantage when it is used against a high bank with little necessity for moving. Most of the power shovels used in placer mining within the State are of the revolving type mounted on caterpillar treads. With this type of locomotion there is no necessity for a pit crew such as is required when the shovel operates on track. Either a Diesel or gasoline engine or electric motor may be used to operate the shovel. When excavating flat - lying gravel deposits the power shovel has considerable mobility, but in working placer deposits in narrow canyons its mobility is restricted. The first cost of installing a power shovel is high and such a shovel will dig only a short distance below the working floor. The height of the bank the shovel will dig is determined by the size of the shovel, the character of the material to be handled, and means taken to insure safety. The horizontal working range from one position is limited by the length of the boom and dipper stick. A power shovel cannot mine selectively, and for this reason it is difficult to follow pay streaks in placer gravels. By itself, the power shovel is not a transportation unit and therefore automobile trucks, conveyer belts, or some other transportation equipment must be used in conjunction with it. In some cases attempts have been made to dispense with transportation by employing portable washing plants, which are dragged on skids or wheels behind the shovel. Standard makes of power shovels have dipper capacities ranging from $\frac{3}{8}$ to 10 cubic yards or more.

The dragline excavator is designed with the same type of chassis and the same housing as those of the power shovel. The principal differences between it and the power shovel are in the design of the superstructure and in a longer boom. The dragline excavator digs with a bucket instead of a dipper; the bucket is dragged along the surface of the ground by cables. Due to its longer boom and the fact that the bucket may be cast some distance beyond the arc described by the boom, it has a greater radius than the power shovel. While a steam shovel stands below the level of the material it is digging, a dragline excavator stands above it, and for this reason may be employed to dig material under water. Because it operates with flexible cables instead of with a rigid dipper stick like the power shovel, it is less positive in its action. The dragline excavator cannot handle rocks as efficiently as the power shovel. Inasmuch as the dragline excavator's digging radius exceeds that of the power shovel, it is more flexible than the latter and it has a certain capacity for selective mining. It cannot be moved rapidly, however, and therefore it is used to best advantage where mobility is not important. Like the power shovel, its initial cost is high. Dragline excavator buckets have a capacity of ⅜ to 10 cubic yards.

The power scraper is a modified tractor- or animal-drawn scraper; the bottomless scraper is dragged over the surface of the ground by means of cables attached to masts or deadmen. The motive power is a stationary double drum hoist driven by Diesel engine, gasoline engine, electric motor, or steam engine. Owing to its cost, steam seldom is used as a source of power in Nevada. The power unit is located usually at the head end of the cable span. The power scraper has a greater range of operation than the foregoing types of excavating equipment. By moving the tail mast, excavation can proceed radially in any direction from the head mast. Within the limits of the cable span the scraper serves both as an excavator and a transporting unit. It is unsuited for digging on hard, rough bedrock. The economic distance of haul is governed chiefly by size of equipment and working conditions. The power scraper can be used in conjunction with automobile trucks, conveyer belts, or another power scraper. The scraper bucket generally is of the self-filling type, so that it can be used to dig either wet or dry material. Digging can be controlled only within narrow limits, so that it is not adapted for selective mining. However, it can be used to remove overburden, and the overburden thus removed may be used to form an inclined runway to transport the pay gravel to the dumping equipment at the head mast. A great variety of power-scraper equipment is manufactured, and the interested reader is referred to makers' catalogs. The initial cost is considerably less than that of a power shovel or dragline excavator. The capacity of a power scraper is from ⅓ to 12 cubic yards.

The slackline cableway is a development of the power scraper. The main difference in operation between the two is that the slackline transports the material through the air on a cable track instead of on the surface of the ground. The bucket used with the slackline is similar in design to the one used on the dragline excavator. The buckets are dumped automatically while suspended from the track cable and are returned to the pit by gravity. By traveling through the air the material is transported at a greater speed than is possible with the power scraper, hence the economic distance of haul is greater. The horizontal range of slackline equipment is greater than that of the power scraper. This excavator is not adapted for digging cemented material or large boulders as the bucket must be loaded in the first attempt to dig. The machine cannot take a number of bites if it fails to secure a full load the first time as can be done by the power shovel. The slackline can be used to dig either wet or dry material, but it is not designed for digging high banks. With the slackline cableway there is less dilution of the pay gravel with barren material because, after the bucket is loaded, it is taken to the discharge point without being dragged over waste material. With additional loading facilities at the head end of the cable span the slackline can be used in connection with other modes of transportation. The capacity of the buckets ranges from $\frac{1}{3}$ to 4 cubic yards.

With the hydraulic giant or monitor, water under pressure is used to excavate, transport, and wash placer gravels in a single operation. The water is conducted to the placer ground in steel pipes of large diameter, and pressure may be obtained either from a natural head or by pumping. When a large volume of water is available at an elevation higher than the placer ground, hydraulicking is the cheapest method of working placer gravels. Because the duty of a miner's inch of water is low, it is uneconomical to pump water for hydraulicking in Nevada. The duty of a miner's inch of water in early California hydraulicking operations was estimated by U. S. Army and California State engineers to vary from 1 to 7.5 cubic yards, averaging about 3 cubic yards. Owing to the cost of pumping water, auxiliary equipment used in connection with the giant, such as either the Ruble, or hydraulic elevator, is not economical in Nevada.

Next to hydraulicking, dredging is the cheapest method of handling placer gravels where conditions are suitable. The usual type of placer dredge consists of a continuous chain of buckets for excavating, a screening and washing plant, and one or more conveyer belts for stacking tailings, all mounted on a scow, the hull of which may be either steel or wood. Several placer deposits in Nevada have been worked in former years by dredging.

The capital cost of a dredge installation is high. Capacity of individual buckets on a dredge is from 3 to 18 cubic feet and daily yardage handled may vary from about 1,000 cubic yards with the smaller dredges to about 9,000 with the larger ones. Gold dredging is a highly specialized branch of mining not within the scope of the present bulletin. The interested reader is referred to bulletins published by the Federal and State Governments and the technical press for additional information.

TRUCKING COSTS IN NEVADA

In working placer deposits in Nevada with mechanical equipment, the cost of transporting gravel to the washing plant is frequently the major item of expense. The cost of truck transportation is governed by so many factors, including road conditions, loading and unloading facilities, type and size of trucks used, length of haul, whether trucks are new or second hand, and the efficiency of drivers, that no general rule can be made. During the course of investigation of mining conditions within the State in 1935 the writer had an opportunity to gather data on the cost of hauling ore with automobile trucks in various mining districts, and these costs have been tabulated in table 4.

The average cost of hauling per ton-mile, as computed from the foregoing table, is 14.8 cents. Virtually all the costs given in the table are the prices for hauling on contract for intermittent shipments of ore. By using company-owned trucks, the cost of haul per ton-mile under average conditions would probably be about 10 cents. Assuming an average of 3,000 pounds of gravel per cubic yard, this price would be 15 cents per cubic yard-mile. With short hauls of less than one mile the cost per ton-mile probably would be increased. As a general rule, however, in estimating costs for placer mining there is a tendency to underestimate rather than overestimate the cost of truck transportation.

PLACER GOLD OCCURRENCES (DISTRICTS) IN NEVADA

Gold placers of actual or potential economic importance are distributed in 14 out of 17 counties in Nevada. As indicated on the accompanying map (plate 1), most of the placers are in the northern and western parts of the State. The more important placers, viewed in the light of past production, are shown by solid circles on the map, while those of lesser importance are indicated with open circles. No doubt other placer deposits occur in the State and several new placer deposits have been discovered in recent years. Much of the placer activity in recent years has

TABLE 4

Cost of Hauling Ore by Automobile Trucks in Nevada

From—	To—	Distance, miles	Cost per ton	Cost per ton-mile	Remarks
Central District	W. P. R. R. siding	4	$1.00	$0.25	Fair desert road. One steep grade. Hand loading.
Rosebud District	Sulphur	8	2.00	.25	Fair desert road. Load from bin.
Rosebud District	Seven Troughs Mill	30	4.00	.13⅓	Fair desert road. Hand loading.
Copper Basin	Battle Mountain	10½	2.00	.19	Do.
Copper Basin	Battle Mountain	9½	1.50	.158	Do.
Galena District	Battle Mountain	15	2.50	.16⅔	Fair desert road. One grade. Hand loading.
Bannock District	Battle Mountain	14	3.00	.214	One steep grade. Hand loading. Road on grade poor.
Copper Canyon	Battle Mountain	18	3.50	.196	Fair desert road. Load from bin.
Cottonwood District	Battle Mountain	10	2.00	.20	One steep grade. Hand loading.
Maysville District	Battle Mountain	18	4.00	.222	Medium grade. Hand loading. Poor loading facilities.
Mill Canyon	Valmy	18	2.00	.111	Fair desert road. Load from bin.
Lewis District	Battle Mountain	12	2.00	.16⅔	Good road. Bin or hand loading.
McCoy District	Battle Mountain	33	6.50	.197	Good road. Hand loading.
Raleigh District	Beowawe	25	3.00	.12	Good road. Load from bin.
Eureka District	E. & P. R. R. siding	4	1.00*	.25	Steep grade. Fair road.
Rhyolite District	Beatty	5	1.25	.25	Good road. Hand loading.
Bullfrog District	Beatty	11	1.25*	.114	Good road. Load from bin.
Chloride Cliff	Carrara	20	3.50	.175	Fair road. Steep grade.
Chloride Cliff	Carrara	24	6.00	.25	Do.
Manhattan District	Tonopah	50	5.50	.11	Good road. One steep grade. Load from bins.
Divide District	Tonopah	6	1.00	.16⅔	Good road. Load from bins.
Copper Contact	Mina	22	2.50	.114	Fair road. Load from bin.
Marietta District	Sodaville mill	12	1.35	.11¼	Fair road. Steep grade.
Simon District	Mina	12	2.25	.18¾	Fair road.
Ashby District	Luning	17	1.75	.103	Do.
Gweenah District	Ledlie	3	1.00	.33⅓	Fair road. Hand loading.
Gold Park District	Austin	44	3.50	.082	Good roads. Hand loading.
Wonder District	Fallon	56	4.50	.084	Excellent highway.
Silver City	Silver City mill	2	.50	.25	Load from bin. Good road.
Silver Peak	Silver Peak mill	11	1.75	.159	Load from bin. Steep grade.

*Prices for doing own hauling; rest are prices for contract hauling.

been confined to the better known districts that were worked in former years. Although the results in some of the old districts have been discouraging, many small - scale placer miners have been able to obtain a living wage by searching for small areas of virgin ground that were overlooked in past operations.

One of the questions frequently asked by the small-scale placer miner is "What are the best placer localities in the State?" This question is difficult to answer because the luck of the individual placer miner is extremely variable. With two men working in the same district, one man may be working on rich placer ground and another, a short distance away, may be having a hard time to make a living. It must also be remembered that most of the best placer ground in the State is held either by location or patent, and the only chance an individual operator with limited capital has to work on ground held by another is under a leasing agreement. Most owners of placer property are averse to leasing to individuals because, it is claimed, that they do not receive their just share of the gold produced.

CHURCHILL COUNTY
EAGLEVILLE DISTRICT

The Eagleville or Hot Springs District is in Churchill County, near the Esmeralda County line, a short distance south of the mining camp of Fairview. A small amount of gold was recovered here from placer operations in 1906. After a short period of prospecting, the placer activity ceased. In 1931, placer gold was discovered in the district in the canyon south of the Eagleville mine. This gold is rough and finely divided and the gravel is angular, indicating that the gold was derived from the gold-bearing veins of the Eagleville mine. In recent years placer activity has been confined to a small amount of desultory prospecting. No important amount of placer gold has been produced in this district.

CLARK COUNTY
ELDORADO DISTRICT

The Eldorado District, also known as the Colorado or Nelson District, is in the Opal Mountains in southern Clark County. The District includes Eldorado Canyon, which drains into the Colorado River. Although the District is credited with a production of several millions of dollars, chiefly in gold and silver, since its discovery in 1857, the production of placer gold has been small. According to old-time residents of the camp, about $6,000 in placer gold was recovered in the early 90's by dry washing operations at the head of Eldorado Canyon. Since that time, the recovery of placer gold has been negligible. In recent years several prospecting shafts have been sunk in the canyon below the town of Nelson in attempts to find pay gravel. This work was abandoned after a short time, so it may be assumed that the results of this work were discouraging.

About 1909 two dredges were constructed to work the bars of the Colorado River two miles above to eight miles below the mouth of the Eldorado Canyon. These dredges were of the suction type. The dredging operations were not successful.

The gold found in the bars along the Colorado River is very fine and is associated with a considerable amount of black sand. The distribution of the gold is erratic and the average grade of these alluvial deposits is too low grade to be worked on a large scale. By working selected portions of the bars and flats along the Colorado River south of Eldorado Canyon, small-scale placer miners recovered about 125 ounces of placer gold in 1934.

MOAPA DISTRICT

In 1931, a few colors of fine gold were panned from the sands in the bottom land of the Muddy River three miles southwest of Moapa. Considerable excitement was aroused over this reported discovery of placer gold and a number of square miles were staked out in claims. Very little prospecting was done and after a short time interest died down.

The gold was found in a sand heavily stained with iron oxide. Intercalated layers of gypsum occur in the sands. The area in which this "placer discovery" was made consists of valley sediments, perhaps hundreds of feet in thickness over a vast area, and in all likelihood the fine gold is so disseminated that the deposits are of no economic importance as far as placer gold is concerned.

GOLD BUTTE DISTRICT

The Gold Butte District is in the south end of the Virgin Range in eastern Clark County, 24 miles southeast of St. Thomas, a station on the Union Pacific System and seven miles east of the Virgin River.

The first discovery of gold in the District was made by Frank Burgess and associates in 1905, in veins. The placer deposits of the District were first worked about 10 years ago by dry washing. The production of gold from veins and placer deposits has been small.

For the past five years A. S. Coleman has been working the placer deposits intermittently on a small scale. The overburden is removed with a team and scraper and the material directly above bedrock is treated in a small, manufactured, portable washing plant. Water for washing the gravel is hauled from Granite Well, about half a mile from the site of operations.

The placer deposits cover a considerable area. They have been formed by the erosion of the veins in the surrounding hills. The depth of gravel varies from 2 to 20 feet, and some angular and rounded boulders are present. For several feet above bedrock the gravel may run up to $2.50 per cubic yard. The gold is invariably fine, and no piece larger than a pinhead has been found. Under the glass, the gold has a nuggety appearance. The gold is associated with a large amount of black sand. The bedrock is altered granite.

In the summer of 1935, three men were employed in working the placer deposits in the district.

On the Colorado River, at a place called Temple Bar, about ten miles southeast from the point where the Virgin River enters the Colorado, a placer deposit has been worked during the fall and winter months by "Slim" Rawlins. The gravel is mined by pick and shovel and the gold recovered by panning. At Temple Bar the banks of the Colorado River are precipitous, and Rawlins lives on a raft moored in the river. Some years ago an attempt was made to hydraulic the gravels by pumping water from the river. The pump was mounted on a raft. This venture was unsuccessful.

The gold is very fine, and a large amount of black sand is present in the panned concentrates. The gravel is cemented.

This placer deposit will eventually be covered by the water of Boulder Lake as the water rises behind Boulder Dam.

DOUGLAS COUNTY
BUCKSKIN DISTRICT

The Buckskin District is in the western foothills at the north end of the Singatze Range, which separates Mason Valley from Smith Valley. Placer gold was discovered in the district in January, 1931, by Fred Hughes. The original discovery was 2½ miles northeast of the old Buckskin mine and about 14 miles northwest of Yerington.

After the initial discovery of placer gold in the District, a large number of claims were located and prospecting disclosed other placer gold occurrences in the vicinity. No definite figures are available on the production of placer gold but the yield is estimated to be in the neighborhood of $9,000.

The original discovery of placer gold was in a small, dry ravine, about one-half mile in length, tributary to Spring Canyon. Rich gravel was found in a shallow depression behind a hard rib of rhyolite tuff crossing the ravine. The depression, about 50 feet long and 10 feet wide, formed a natural concentration area. Lying on the rough rhyolite bedrock is a stratum of brown clay mixed with small angular fragments of rock. This clay is covered with less than a foot of unconsolidated sand and grit. Some of the clayey material yielded $1 in gold per pan. Several hundred dollars were taken from this area.

In 1932, the upper part of the ravine was leased to C. F. Scott and Sam C. Case, who produced about $1,000 by working on a small scale for a short time. In working the ground, the lessees removed the soil and gravel above bedrock with a Fresno scraper and team of horses. The average depth of the overburden was four feet. After stripping the overburden, the decomposed rhyolite bedrock was ploughed to a depth of 10 inches and loaded by means of horse - drawn scraper and dumping platform into a small automobile truck. The material was hauled to Artesia, about 3½ miles distant, where water is available. The treatment plant comprised a power - driven sheet - iron shaker box equipped with ¾ - inch and 8 - mesh screens and a sluice box 10 inches wide, 8 inches deep, and 48 feet in length constructed of 2-inch plank. Coarse woven-wire screen was used for riffles, and only the minus 8 - mesh product was sluiced. The material treated is reported to have averaged better than $10 per yard. The cost of treating was $1.50 per cubic yard.

In 1933, this placer was taken over by a stock company called the Ambassador Gold Mines, Ltd. The company went to considerable expense to install equipment for working the placer by hydraulicking. An iron pipe line 8 and 12 inches in diameter, joined with Victaulic couplings, was laid from Artesia to the placer ground, a distance of about four miles. Water for hydraulicking is obtained from several artesian wells in the valley. Electrically driven pumps were installed at the wells to lift the water to the placer, the difference in elevation between the two points being several hundred feet. Pumping capacity is 4,500 gallons per minute.

Sluice boxes are 30 by 30 inches in section and 1,400 feet in length. The bottom of the sluice boxes is lined with ⅛-inch steel plate, upon which 30-pound rails are placed transversely for riffles. At the lower end of the sluice an undercurrent of special design is used to catch the fine gold not caught in the riffles.

The placer was operated in 1934 and for a short time in 1935.

3

Two 5 - inch hydraulic giants were employed. The site of the hydraulic operations is shown in figure 17. It is stated that the company became involved with ranchers in the valley over water rights and was forced to discontinue operations in 1934.

The gold is both fine and coarse; the largest nuggets found had a value of about $2 each. A considerable amount of black sand is present. The depth to bedrock varies from two feet at the upper end of the ravine to 16 feet at the lower end. The placer material is largely rhyolite fragments, with no fragments larger than cobblestones. Occasional fragments of silicified wood

Figure 17. Site of hydraulic operations at Ambassador placer, Buckskin District.

are found. The bedrock is uneven, altered rhyolite considerably fractured.

About one mile southeast of the Ambassador property, on the alluvial fan sloping toward Artesia, is the Guild-Bovard placer. This placer was worked in 1933 and part of 1934 by eastern interests. The alluvium was mined by power shovel and dumped directly into the hopper of the portable washing plant shown in figure 18. From the hopper, the gravel was discharged into a combination washing and screening trommel 20 feet long and 4 feet in diameter. The washing section is 14 feet in length and the screening section 6 feet. Between the two sections is a retarding ring 3 inches high. Trommel screen is ¼-mesh woven wire. The minus ¼-mesh product passes over two inclined riffle decks arranged in series, each deck 4 feet wide and 12 feet long with angle-iron riffles 2½ by 1½ inches spaced 4½ inches apart. An Akins classifier, 30 inches in diameter and 11 feet long is used for dewatering the sluice tailings. The overflow from the classifier goes to a semicylindrical tank 8 feet long and 5 feet in diameter. Water from the settling tank is pumped back to the trommel.

Power for the washing plant is provided by gas engines mounted in the frame. The plant is moved on wheels which run in 3-by-12-inch channel iron.

Several difficulties were encountered in the operation of the

Figure 18. Washing plant at Guild-Bovard placer, Buckskin District.

plant, namely, the difficulty of disposing of the tailings, the character of the gravel, which is tightly cemented, and the limited depth to which the power shovel could dig and dump directly into the hopper of the machine. The maximum digging depth was about 18 feet.

The water supply for the placer plant is obtained from a well 100 feet deep located in Cement Canyon, two miles from the placer. This well has a flow of several hundred gallons per minute. Water is pumped from the well by a deep-well pump connected to a Buick automobile engine. A 4-inch steel pipe line carried the water by gravity to the washing plant.

In 1934 the ground was partly sampled by Keystone drill. Sixteen holes from 9 to 76.5 feet deep were drilled to bedrock. The results of this sampling are shown in the following table:

TABLE 5
Results of Drilling at Guild-Bovard Placer, Buckskin District

Hole No.	Depth of pay gravel	Average value pay gravel per cubic yard
1	5.0 feet to 44.0 feet	11.0 cents
2	5.0 feet to 32.6 feet	5.0 cents
3	3.0 feet to 35.0 feet	22.2 cents
4	4.5 feet to 49.0 feet	40.1 cents
5	4.5 feet to 9.0 feet	21.5 cents
6	3.5 feet to 41.5 feet	(*)
7	3.5 feet to 21.5 feet	21.8 cents
8	4.5 feet to 28.0 feet	15.3 cents
9	4.0 feet to 9.0 feet	103.2 cents
10	4.5 feet to 14.5 feet	73.8 cents
11	3.0 feet to 53.0 feet	(*)
12	4.0 feet to 7.5 feet	15.2 cents
13	4.0 feet to 12.0 feet	29.0 cents
14	43.0 feet to 73.5 feet	†72.1 cents
15	73.0 feet to 76.5 feet	13.8 cents
16	4.0 feet to 53.0 feet	8.5 cents

*Trace.　†Water encountered at 60 feet.

The foregoing values per cubic yard are based on a gold price of $34.95 per fine ounce. The best values are concentrated near bedrock. The alluvium is composed of rhyolite gravel and fragments, sand, and some clay. Bedrock is rhyolite. The material is tightly cemented so that in drift mining blasting would be necessary. The gold may be classified as shot gold. Some black sand is associated with the gold. The largest nugget found had a value of $5.20 at the current price. The fineness of the gold ranges from 780 to 820.

GENOA DISTRICT
The Genoa District lies upon the east slope of the Sierra Nevada Range in Douglas County, a short distance west of the old town of Genoa. The District was organized in 1860 and a large amount of development prospecting was done on the lode deposits with discouraging results. Deposits of Tertiary gravels occur in the District, but they are of doubtful economic importance. A small amount of placer gold was produced in prospecting the deposits in 1916. In recent years there has been no placer activity in the District.

MOUNT SIEGEL DISTRICT
The Mount Siegel placer District lies on the north side of Mount Siegel (elevation 11,000 feet) in east Douglas County, about 20 miles east of Minden, Nevada. The placers are in the Pine Nut Range at an elevation of 7,100 feet above sea level. The first discovery of placer gold in the District was in 1891. The Ancient Gold Placer Mines Company, formerly known as the Buckeye Placers, owns most of the placer ground in the District. This company's holdings comprise 15 quarter-section

claims, totaling 2,440 acres. George Slater, of Minden, is the principal owner of the company. For many years the placers have been worked on a small scale by sluicing, rocking, and panning during the early spring as long as water from melting snow would last. The District has been handicapped by lack of ample water in the immediate vicinity. The production of placer gold, based largely on statements of previous owners, has been a little over $100,000.

The placer deposits occur in a large depression in the granite mass comprising the Pine Nut Range. Within this depression the surface has been crevassed by recent water courses making an undulating relief of hills and ravines. Gold is scattered over a considerable area, with some concentration in the ravines or on a false bedrock of pipe clay or hardpan. The gravels consist of loose, unassorted rock, gravel and sand, and soil, with some large boulders. Gold, both fine and coarse, as well as nuggets, is disseminated through the gravels. The largest nugget found had a value of $250; another nugget had a value of $168. Fineness of the gold averages about 880. Some of the gold, it is believed, has been reconcentrated from gravels deposited in a Tertiary river system that coursed easterly from the Sierra Nevada Range, while some has been derived from the erosion of gold - bearing quartz veins and stringers in the Pine Nut Range.

Development of the property has, to a large extent, been confined to Pinto, Dudley, and Black Horse Gulches. Development consists of numerous shafts ranging from 8 to 170 feet deep, several tunnels, and some open cuts. The bedrock, which is presumably granite, has not been prospected to any extent.

In the middle nineties a considerable amount of money was spent to bring in water from a small lake formed in a depression above the placer ground. This depression, about 1,000 feet long, 300 feet wide, and 30 feet deep formed a catch basin for water from melting snow. This natural reservoir was tapped by a tunnel 700 feet in length. A 22-inch pipe with gate valve was installed at the inner end of this tunnel for controlling the flow of water. The 22-inch pipe carried the water from the reservoir to a ditch from whence it was diverted into 1,700 feet of 15-inch pipe for delivery to the hydraulic giant. The water was under a head of 190 feet.

In 1896, when this reservoir was not more than a quarter full of water, the company had a six-day run, during which period about 4,000 yards of gravel were moved by the giant, but only about 600 yards were cleaned in the sluices. The owners were compelled to suspend operations because the water supply was exhausted. From the 600 yards of gravel sluiced it is reported that the clean-up contained 75.14 ounces of gold which netted the owners $1,322.80; in addition nuggets to the value of $70 were recovered. Allowing for the uncertainty as to the exact yardage treated, this would be an average of about $2 per cubic yard of gravel treated at the old price of $20.67 per ounce for gold.

As a result of this undertaking it was demonstrated that it was impossible to procure sufficient water in the immediate neighborhood for successful hydraulicking operations.

Some years later an attempt was made to work the placers on a larger scale than by hand methods by installing a washing plant and excavating the gravels with a Bucyrus shovel run by

Figure 19. Site of former placer mining activity on ground owned by Ancient Gold Placer Mines Company, Mount Siegel District.

a Diesel engine. (See figure 19.) It is reported that this venture was unsuccessful because the washing plant did not make a good saving of the fine gold associated with black sand. It is stated that 1,200 cubic yards were treated and $300 in gold was recovered.

It is difficult to estimate the average gold content per cubic yard of the gravels because of the occurrence of fine gold, coarse gold and nuggets, with erratic distribution of values. If the results of the past operations are taken as a criterion, a large yardage of auriferous gravel is indicated. Further sampling is necessary to determine the average gold content per cubic yard. Reports by engineers who examined the property in the early days are encouraging. If sampling results justify working the property on a large scale, the future of the property will depend largely on the amount of water that can be obtained at reasonable cost.

ELKO COUNTY
ALDER DISTRICT

The Alder or Tennessee Gulch District is in north central Elko County, 10 miles south of Rowland. Considerable placer mining was done in the early days, as is indicated by the old placer tailings on Gold Run Creek, half a mile north of Baker Ranch. In recent years a small amount of placer prospecting has been done in this area, but the results of this work have not been encouraging.

AURA DISTRICT

The Aura District, also known as the Bull Run, Centennial, or Columbia District, is in the central part of the Bull Run Mountains, 20 miles southwest of Mountain City in northern Elko County. The District was discovered by Jesse Cope and party in 1869. The placer deposits, which have been of secondary importance to the lodes, were first worked in the seventies. In 1905, a company was organized to work the gravels of Bull Run Basin on a large scale. An extensive system of ditches, flumes, and pipe lines was built to hydraulic the gravels, but after washing only a few hundred yards the project was abandoned. It is said that some gold was recovered, but, owing to the size of boulders, exploitation of the gravels was not profitable.

In 1925, a ditch was constructed to bring water to the placers from Blue Jacket Canyon for the purpose of hydraulicking. This venture also was not successful and operations ceased after a short time. In recent years the placer deposits have not been worked.

CHARLESTON DISTRICT

The Charleston District, also known as the Copper Mountain or Cornwall District, is in northern Elko County, about 55 miles north of Deeth. It is on the south side of the Jarbidge Mountains, and in particular on the south side of Copper Mountain, which rises to a height of over 10,000 feet above sea level. Charleston came into existence in 1876 when placer gold was discovered four miles north of the townsite on 76 Creek near the base of Copper Mountain. The bed of 76 Creek produced considerable gold in the years following the discovery of placer gold. Since water is available in 76 Creek the placer deposits were worked thoroughly. Lying between 76 Creek and the Bruneau River are Pennsylvania Gulch, Union Gulch, Dry Ravine, and Badger Creek, in which placer deposits also were worked in the early days.

The placer gravels of the Charleston District extend for miles along the Bruneau River.   They consist mainly of well-rounded pebbles derived from rather coarsely crystalline rhyolite, which is one of the principal rocks of the District, and smaller amounts of quartzite and granite pebbles.   In places these gravels are 50 feet thick, and often rest upon a light yellow clay of decomposed tuff, which may indicate the bottom of the gravel beds. A long period of erosion and working over by earlier streams is indicated by the well-rounded character of the gravels.   The placers of the Mountain City District, 35 miles to the northeast, appear to be of similar origin.

In 1907, a Utah company, at a cost of $25,000, built a ditch several miles in length around the south slope of the mountains in order to bring water from a tributary of 76 Creek for use in hydraulicking the gravels in the Badger Creek area.   The tributary normally carries but little water, and it is reported that water flowed through the ditch for a short time only during one season.

In the spring of 1932, a Denver group obtained an option on the Prunty Ranch, which is traversed by the Bruneau River and Badger Creek.   Considerable prospecting was done and it was estimated that the gravel contained an average of 75 cents worth of gold per cubic yard.   For some unknown reason this prospecting was not followed by any extensive development of the placer ground.

In recent years, during the spring and summer seasons, from 5 to 30 men have been engaged in placer mining in Pennsylvania, Union, and Dry Gulches.   Most of this work is done by small-scale sluicing when water is available.   One of these sluicing operations is shown in figure 20.   The average yield per man by such small-scale operations is less than wages.   The gold recovered is quite fine.

GOLD BASIN DISTRICT

The Gold Basin or Rowland District is in north central Elko County, about four miles from the Idaho-Nevada boundary line.

Small-scale placer operations have been carried on intermittently in the District for a number of years.   In 1931, A. S. Longwill treated a small amount of gravel on the north fork of the Bruneau River.   The gravel is reported to have yielded less than $1 per cubic yard.   Water for sluicing was impounded in a small reservoir.   The District has never produced any appreciable amount of placer gold.

ISLAND MOUNTAIN DISTRICT

The Island Mountain or Gold Creek District is at Island Mountain in the vicinity of Gold Creek in north Elko County, 75 miles north of Elko and about 25 miles south of the Idaho-Nevada State line. The District derives its name from an isolated mountain that rises over 1,000 feet above the surrounding terrain. The average elevation of the District is nearly 8,000 feet above sea level.

The placers of Island Mountain were discovered by Penrod, Rouselle, and Newton in 1873. They soon became one of the most prominent placer areas in the State and attained a large

Figure 20. Sluicing near the mouth of Union Gulch, Charleston District.

production. The first operations were confined largely to sluicing, but in subsequent years a large investment was made in ditches and a pipe line to bring in water for the operation of hydraulic giants. A ditch was constructed from the Owyhee River for a distance of five miles and 2,500 feet of pipe line installed, which brought the water to the placer ground under a head of 300 feet. Water for hydraulicking was available for only a few months each year, but considerable gold was recovered. In addition, a large amount of panning and rocking was done by individuals in the early days. According to an early report of the State mineralogist for Nevada,[13] a number of Chinese and Americans worked the claims on Hope Gulch

[13] Stretch, R. H., Biennial Report of the State Mineralogist, State of Nevada, 1873, pp. 27, 28.

and recovered as high as $2.50 per pan and as much as $30 per day per man with rockers.

Until recently, the Island Mountain placers have not been worked since the early days. In 1934, a group from Lodi, Calif., obtained a bond and lease on three placer claims near Gold Creek owned by Fred Robertson of Boise, Idaho. A 4-inch pipe line, sluice, and two caterpillar tractors with scrapers were installed to work the placer gravels. Water was obtained by impounding in a reservoir the water in Gold Creek, and it was pumped to the head of the sluice by a centrifugal pump driven by a gasoline engine.

The gold occurs in the gravels near the surface, the average depth worked by caterpillars and scrapers being about seven feet. Some boulders and clay are present in the gravels. The gold recovered is fairly coarse. In 1935, the placers were worked for a period of six months after which the water supply diminished to such a degree that it was necessary to suspend operations. Eight men were employed on this work and the results are reported to have been profitable.

Some boulders and clay are present in the alluvium. The gold recovered is fairly coarse. The source of the gold is the quartz veins in the vicinity and at the head of Gold Creek.

Frank McGregor and associates from Ogden, Utah, have acquired extensive placer holdings in the Island Mountain District. This ground was located several years ago. The only patented ground in the area is the Penrod group of three locations owned by D. A. Duryee and others, of Everett, Wash.

MOUNTAIN CITY DISTRICT

The Mountain City District, also known as the Cope District, is in the northeastern part of the Centennial Range on the north fork of the Owyhee River, about one and one-half miles east of the Duck Valley Indian Reservation. Silver-gold deposits were discovered in the District in 1869 by Louis Cope and others, who were on their way from Silver City, Idaho, to the White Pine District, Nevada. Following the discovery of lode deposits, placer gold was found in the Owyhee River, but little attention was paid to it at the time. Compared with the lode deposits, the placers are relatively unimportant.

In the middle seventies, placer gold was discovered in Grasshopper Gulch, north of Sugar Loaf Peak. The gulch was profitably mined for half a mile. Many piles of tailings are evidence of the activity of the early placer miners. Stretch, in the biennial report on the mining activity in the State in 1875–1876,[14] stated that the placer mines on both sides of the Owyhee River yielded well. From 1870 to 1890 much of the work was done by R. M. Woodward and Alley Harris and associates, who constructed a ditch from Mill Creek and hydraulicked part of the ground.

In recent years small-scale placer operations have been carried on intermittently in Hansen Gulch, which is a tributary to Grasshopper Gulch, and along the Owyhee River several miles north

Figure 21. Hydraulic mining on Van Duzer Creek, Elko County, in 1894.

of Mountain City. The production of placer gold from these operations has been small.

VAN DUZER DISTRICT

The Van Duzer District joins the Mountain City District on the north. Van Duzer Creek, which traverses the District, is a small stream that flows eastward and joins the north fork of the Owyhee about six miles south of Mountain City. Placer gold was discovered in the District in 1893 by Rutley M. Woodward. No accurate data are available on the production of placer gold here, but from information gathered from several sources it is estimated to have had a value of about $100,000. Figure 21 is a view of hydraulicking operations in the District in 1894.

---

[14]Stretch, R. H., Biennial Report of the State Mineralogist, State of Nevada. 1875–1876, p. 25.

Portions of the main fork of Van Duzer Creek, for a distance of several miles, have been washed for placer gold. Woodward is reported to have taken out gold to the value of $50,000 in the years following the first discovery. The ground was sold subsequently to Alley Harris and others who worked it until 1910. Harris reported that the placer paid him an average of $5 per linear foot of creek bed for a distance of 7,500 feet, a total of $37,500. The first placer mining was done by hand shoveling into sluice boxes. Later, several hydraulic monitors were employed. Two small reservoirs were constructed in Van Duzer Creek and 10-inch steel pipe lines were laid to supply water for the monitors.

The alluvium is composed of fine gravel and subangular and well - rounded pebbles. The bedrock is presumably limestone. The gold varies from fine dust to nuggets weighing five or six ounces. The fineness of the gold averages about 820. The source of the gold is probably from the quartz veins at the head of the stream. Perhaps some of the gold was derived from the ancient fluviatile gravels that cover large areas in northern Elko County. The average depth of the gravels is about 10 feet.

Inasmuch as the best portions of the placer area have been worked, there has been only a small amount of placer activity in the District in recent years. In 1932, the placer ground was leased to a Los Angeles operator who employed four men. They shoveled the gravel onto a conveyer belt, which discharged into a trommel. The trommel oversize was rejected and the under-size diverted into a sluice. The amount of gold recovered from this work was small. The stream is perennial so that small-scale sluicing operations can be carried on continuously when the weather is favorable.

TUSCARORA DISTRICT

The Tuscarora District is in Elko County on the southeastern slope of Mount Blitzen in the Independence Range, about 50 miles northwest of Elko. The gold placers were discovered in 1867 by the Beard brothers, and several years later the rich silver and gold veins were found. The placer deposits were worked extensively for many years, first by Americans and later by Chinese placer miners. Complete data on the production of placer gold are not available, but, according to Emmons,[15] the placers are reported to have yielded $7,000,000. An interesting sidelight on placer gold production at Tuscarora was given to the writer by Roy L. Primeaux who is familiar with the history

---

[15] Emmons, William H., Some Mining Camps in Elko, Lander, and Eureka Counties, Nevada: U. S. Geol. Survey Bull. 408, 1910, p. 59.

of the camp over a period of many years. According to Mr. Primeaux, the Wells Fargo express rate on gold was so high in the early days that much of the gold was consigned to San Francisco by bullion buyers as hardware in order to avoid the high express rate.

In the same year that the Beard brothers made their discovery of placer gold at Tuscarora, news of the strike reached Austin, Nevada, and over 100 men left Austin in a band that year, bound for the new strike. The band was well equipped and armed against hostile Indians, who made prospecting in Nevada both adventurous and dangerous at that time. Many of these pioneers remained at Tuscarora and found the placer diggings profitable. J. Ross Browne, a distinguished mining geologist of the time, described the placers as being three miles long, five feet deep, and rather narrow. He predicted that a score of men could make from $10 to $20 per day by hand washing the gravels.

At first the deposits were worked by Americans principally by ground sluicing. Water for placer operations was supplied through two ditches constructed in 1868 and 1869 with Chinese labor. The Beard brothers' ditch was four miles long and brought water to the diggings from McCann Creek; the other ditch was six miles in length and conveyed water from Three Mile and Six Mile Creeks. In 1900, these pioneer ditches were abandoned and the water rights acquired by ranchers.

In ground sluicing in the early days, horn silver and some native silver was found in the clean-ups; this fact led the early prospectors to the rich silver veins in the vicinity, which became large producers. Rich silver float also was discovered, and for five or six years in the seventies, the Winslow 10 - stamp mill operated solely on float ore.

In the seventies, the Beard brothers and others leased their placer holdings to Chinese on a royalty basis of 10 percent of the bullion recovered. According to Mr. Primeaux, at one time there were about 2,500 Chinese at work in the diggings. The Chinese produced from $2 to $15 per day per man.[16] The Chinese worked only during the spring and summer months when water was available. During the fall and winter they cut sagebrush for the steam plants in the nearby silver mills. Yen Tin, one of the last survivors of the Chinese placer miners, died at Tuscarora in 1927. In 1934, several boys playing in the camp discovered a cache of gold dust and nuggets, valued at $1,200, that had been hid by Yen Tin near his cabin.

[16] Whitehill, H. R., Biennial Report of the State Mineralogist, State of Nevada, 1871–1872, p. 24.

A surprising feature of the early placer operations is the relatively small area worked in proportion to the production of $7,000,000 with which the Tuscarora placers are credited. The workings are confined to the gullies on the gently sloping sagebrush-covered hills bordering the west side of Independence Valley. Typical old workings, now almost obscured by cloudburst action, are shown in figure 22. Eureka Gulch and Gardner Ravine are known to have been very rich. The flat, shallow ravines draining south into Independence Valley also yielded well.

The depth of the gravel worked in former operations varied from four to ten feet. The gold occurs as dust and nuggets largely concentrated on bedrock. The largest nugget found in

Figure 22. Placer tailings from old workings in the Tuscarora District.

the District weighed nine pounds, although it contained considerable quartz. Nuggets weighing up to one ounce were common. The bedrock is largely rhyolite, which is the principal rock of the District. The source of the gold is presumably the gold veins that occur in rhyolite and andesite north and west of the diggings.

After 1900 only small sporadic placer operations were carried on. In recent years a small amount of gold has been recovered by individuals. The Harris claims, one and one-half miles west of Tuscarora, have been worked for a number of years by an operator who uses a Chinese rocker. In 1931, a group of men sluiced the gravel in Review Gulch, a short distance west of town. Water was obtained from several springs in the vicinity.

This work was confined to the lower part of the gulch in an area that may have been overlooked by the early placer miners. In the same year some placer mining also was done in Stovepipe Gulch, water for sluicing being obtained from the old Stovepipe mine shaft. In 1934 and 1935, a few itinerant placer miners prospected the District for virgin areas rich enough to be worked by hand methods, but the results of this work were not encouraging.

A United States Geological Survey report by Emmons, in 1910,[17] states: "A large acreage of ground west of Tuscarora has been located and sampled with drills. It is said that much of this ground will pay to work with dredges, and two companies are planning such operations. A large number of samples are reported to have given an average of about 14 cents (based on gold value of $20.67 per ounce) per cubic yard."

No dredge has operated in the District, and, if the above information can be checked by additional sampling and the other factors are favorable, this ground may offer interesting dredging possibilities. Perhaps sufficient water for dredging purposes can be obtained from the old Dexter gold mine at Tuscarora, now flooded. A large volume of water was encountered in the underground workings of this mine when it was worked in the early days.

ESMERALDA COUNTY
KLONDYKE DISTRICT

The Klondyke District is 14 miles south of Tonopah in the southern Klondyke hills in east Esmeralda County, near the Nye County border. The lode deposits in the district were discovered by Court and Bell in 1899, but it is said that some Chinese placer miners were active in the District in the middle seventies. For several years following the discovery of the lode deposits, some placer mining was carried on by dry-washing methods. One nugget valued at $1,200 is reported to have been found. In general, the results of the placer mining were not very profitable; in recent years there has been no placer mining in the District.

TULE CANYON

The Tule Canyon District is ten miles south of Lida in south Esmeralda County at the southern end of the Silver Peak Range. Tule Canyon is tributary to the north end of Death Valley. Placer gold was discovered in the canyon in 1876, although it is said that these placers were worked by Mexicans prior to

---

[17] Emmons, William H., A Reconnaissance of Some Mining Camps in Elko, Lander, and Eureka Counties, Nevada: U. S. Geol. Survey Bull. 408, 1910, p. 62.

1848, while Nevada was part of Mexico. After 1876, a number of Chinese placer miners worked in the District. The old workings in Nugget Gulch and at the head of Tule Canyon indicate that a considerable amount of placer mining, mainly drifting along bedrock, was done in former years. Reminders of Chinese occupation have been found in the form of neck yokes, which the Chinese employ to transport materials. A report by the State mineralogist for Nevada, written in the late seventies, mentions that chispas (Mexican word for nuggets) worth up to $50 were found in the District.

Placer gold has been found in the District over an area ten miles square, but the best ground is in Tule Canyon and the side gulches tributary to it. Owing to the scarcity of water in the immediate vicinity, the placer deposits have been worked, until recent years, solely by hand methods. The only water available in the vicinity is derived from a few small springs.

In 1933, the Los Angeles Rock and Gravel Corporation started to work placer ground two miles in length comprising eight claims, at the upper end of Tule Canyon. This company has operated in the District for the last three years whenever conditions were favorable. In 1935, the placer season started on April 10 and the plant closed down for the year on October 29 because of freezing weather.

The alluvium, which varies from 12 to 18 feet deep, is stripped to three to six feet above bedrock with a ½-cubic-yard Northwestern shovel equipped with dragline bucket. The usual procedure is to remove overburden on two shifts of eight hours each and mine the material near bedrock on the other shift. The capacity of the washing plant is 70 cubic yards per shift and 12 men are required to operate the placer layout.

The pay gravel is loaded by the power shovel into Ford dump trucks and hauled about a quarter mile to a point in the canyon opposite the treatment plant shown in figure 23. The gravel is dumped into the canyon, whence it is picked up by slackline cableway bucket and conveyed to the washing plant.

The treatment plant consists of a grizzly over the hopper above the trommel. The trommel is 12 feet long and 3 feet in diameter and has a punched plate screen with holes ¾ inch in diameter. From the hopper the gravel is fed to the trommel by eccentric shaker feeder. The plus ¾-inch product constitutes about 50 percent of the material treated. The trommel oversize is discharged onto a conveyer belt 100 feet long and 18 inches wide, which carries it to the waste dump. The washing plant is operated by a model A Ford engine.

The trommel undersize is diverted to a sluice box 80 feet long, 2 feet wide, and 1 foot high, made of 2-inch plank. Slope of sluice is ¾ inch per foot. Rock riffles to a depth of about 3 inches are used to catch the gold. About 90 percent of the gold is recovered in the first 15 feet of the sluice.

Water for washing the gravel is piped by gravity from springs in Weatherspoon Canyon through 6,000 feet of 1½- and 2½-inch iron pipe. Flow from the springs is about 18 gallons per minute. The sluice tailings discharge into a series of three settling ponds in the canyon below the washing plant. After settling, the water is returned to the trommel by a 3-inch centrifugal pump driven by a 20-hp. Fairbanks-Morse engine. Approximately 75 percent of the water is recovered in each washing cycle.

The alluvium is composed of sand, gravel, and medium-size

Figure 23. Washing plant and coarse tailings dump, Tule Canyon District.

water-worn boulders. Some large boulders are present, but in less proportion than in the lower part of the canyon. The bedrock is granite and porphyry. Some cemented gravel is present above the paystreak, but blasting is not necessary in excavating with a power shovel. Gold is fairly coarse and rounded. Fineness of the gold is about 650. The largest nugget found in 1935 had a value of $14. The source of the gold is presumably the Eagle and other lode mines in the vicinity.

In Tule Canyon, about one and one-half miles below the property worked by the Los Angeles Rock and Gravel Corporation, E. E. Layton of Goldfield, Nevada, was installing placer equipment in the fall of 1935 to work the Ray White placer holdings, comprising three claims and covering about one mile of the canyon. Placer equipment on the ground at the time of visit included a 1-cubic-yard-capacity Bucyrus shovel run by a Diesel engine, a dragline scraper operated by a double drum gasoline hoist, and a washing plant. The plan of operations consisted in stripping the overburden to a depth of ten feet with the power shovel and transporting the six feet of pay gravel above bedrock to the washing plant with a dragline scraper.

The washing plant consists of a trommel and shaker sluice that operates with a longitudinal motion. Above the trommel is a hopper equipped with a grizzly over the top, and a Challenge feeder for feeding the grizzly undersize into the trommel. The trommel is 3 feet in diameter and 10 feet long with $\frac{1}{8}$- and $\frac{1}{2}$ - inch screens. The shaker sluice is 16 feet long, 18 inches wide, and 4 inches deep, made of 2-inch plank. Riffles in the sluice are 1 x 3-inch boards spaced 1 foot apart. The upper edges of the riffles are covered with narrow strips of strap iron to lessen the wear.

Water for the washing plant will be obtained from a well 16 feet deep, which taps a subterranean flow in the canyon a short distance above the washing plant. In a pumping test, a 3,000-gallon tank was filled from this well in two hours.

The average depth to bedrock in the canyon is 16 feet. Four shafts have been sunk in the canyon and drifts extended from these shafts along the granite bedrock. Preliminary sampling in these workings indicated that the value of six feet of the material above bedrock will average $2 per cubic yard. The average width of the canyon is approximately 100 feet. The alluvium is poorly sorted and consists of angular and rounded boulders in wash gravel. Many large granite boulders are present and these, no doubt, will prove troublesome in mining. In places the gravel is slightly cemented. The gold is coarse and rough and some is attached to quartz, indicating that it has not traveled a great distance from its original source. The fineness of the gold is reported to be about 800.

PIGEON SPRINGS DISTRICT

The Pigeon Springs or Palmetto District is ten miles west of Lida in the Sylvania Mountains. Lode deposits in the District were discovered in 1866. The placer ground covers several square miles near the California-Nevada boundary line. From the viewpoint of past activity the placers are of minor importance as compared with the lode deposits. The early activities on the placers gave good results with small-scale dry-washing equipment. The greatest activity on the placers occurred in 1914, and since that time the District has produced a small amount of placer gold. In recent years a small amount of work has been done on the Rigsby placer. The water resources of the area are limited, so that placer operations have been confined to dry washing.

TOKOP DISTRICT

The Tokop District, also known as the Gold Mountain or Oriental Wash District, is in south Esmeralda County, 29 miles south of Goldfield. Oriental Wash is a small valley filled with desert gravels lying between the Slate Ridge on the north and Gold Mountain on the south. From Tokop, this valley drains westward into the north end of Death Valley. The District was discovered in 1866 and attracted considerable attention when, in 1871, rich gold veins in the Oriental mine were discovered. The Oriental was one of the richest properties in the District. Upon the surface many tons of ore in the form of float were picked up and sent to Eureka, Nevada, for treatment. A small amount of placer gold was produced in the District in past years, but in recent years there has been no placer mining of any consequence here.

EUREKA COUNTY
LYNN DISTRICT

The Lynn District is about 20 miles northwest of Carlin in the Tuscarora Range in north Eureka County. The placer deposits were discovered in 1907 by Joe Lynn and, following their discovery, a brief boom ensued. Since their discovery the placers have been worked intermittently by small-scale methods. The total production of gold is estimated to have been about $140,000; the bulk of this was from the placers.

The Lynn District covers a large area, and placer deposits have been worked in Lynn, Simon, Rodeo, and Sheep Creeks, which cut the high, well-rounded, sagebrush-covered hills. The formation in the District is largely bedded rhyolite flows traversed by a number of small quartz veins and stringers in shattered zones. The erosion of the veins and stringers has supplied the material for the placer deposits. The gravels are fairly coarse and contain a large proportion of medium-size subangular boulders. In general, the placers are rich but rather narrow and thin and in places are covered by considerable overburden. The richest gravels are at the upper ends of the ravines, where the gold is coarse and rough. In the lower portions of the ravines the gold is much finer and is associated with a considerable amount of black sand. Bismuth oxide and sulphide have been found in the placers. The Lynn placers contain perhaps the purest gold of any of the placer deposits in the State; the fineness ranges from 920 to 960.

During the last five years the placers have been quite active, particularly during the spring when surface water for washing the gravels is available. In 1935, about 12 small-scale placer operations, employing 30 men, were working in various parts of the District. Of these, nine used side - shake power - operated sluice boxes, two used dry washers, and one used a power-operated shaking table of special design. On the average, in 1935, the operators netted better than wages for the time they worked. The most promising ground in the District has been

Figure 24. Portable placer plant, Lynn District.

located for many years, so that some of the operations are conducted on a leasing basis, the customary royalty being 10 percent of the gross bullion returns.

In 1935, C. A. Clemans, owner of a placer claim located at the lower end of Lynn Creek Canyon, worked his ground with a power shovel and portable washing plant, shown in figure 24. The shovel is a home-made affair built by Clemans in the winter of 1934 – 35 at Portola, Calif. It is mounted on caterpillar tractor treads that have been lengthened several feet. The driving mechanism for the treads is the rear end of a 4-ton Locomobile automobile truck. The shovel power plant is a 6-cylinder Star automobile engine. The turntable is made of 40-pound rails bent to form a circle. Capacity of the dipper is one third of a cubic yard, and from 100 to 300 yards can be handled in ten hours, depending upon the character of material dug. Fuel consumption is one gallon of gasoline per hour and one quart of oil per day. Cost of shoveling is approximately 8 cents per cubic yard in average ground. The weight of the shovel is 7¾ tons, and its cost was about $2,000, including labor. The only new parts in the shovel are the sprockets for the chain drives, and these were made by Clemans.

The washing plant, also designed and built by Clemans, comprises a hopper, trommel, tailings stacker, and two side-shake sluice boxes. It is capable of washing about six yards per hour, and is operated by a 4-cylinder Whippet automobile engine. Fuel consumption is six gallons of gasoline and one quart of oil per eight-hour shift. The plant is mounted on 6 x 8-inch timber skids shod with steel plates, so that it can be dragged into position by the dipper of the power shovel. Only two men are required to operate the entire plant.

The shovel dumps directly into a 1-cubic-yard hopper above the trommel. Above the hopper is a grizzly made of 20-pound rails with 3-inch openings. From the hopper, the material is fed into the trommel, 8 feet long and 3 feet in diameter, mounted on rollers and driven at a speed of 20 r.p.m. by a chain drive. The slope of the trommel is 3 inches in 8 feet. The screen is made of sheet iron into which slots have been cut with an acetylene torch; the slots, which are staggered, are 6 inches long and ⅜ inch wide, spaced at intervals of 4 inches around the circumference. A woven-wire screen was tried but, owing to the character of the material treated, it was found that the water ran through it too rapidly, with a consequent decrease in the disintegration of the alluvium. The minus ⅜-inch product from the trommel constitutes about 30 percent of gravel. The trommel oversize is discharged by gravity into the boot of the belt-and-bucket stacker. The stacker boom is 20 feet long and steel buckets are 7 inches wide by 12 inches long.

The trommel undersize is diverted to two side-shake sluice boxes placed side by side. Each box is 12 feet long, 13 inches wide, and 10 inches deep. Longitudinal riffles are used in the upper 3 feet of the boxes and transverse riffles are used for the remaining 9 feet. Riffles are made of strap iron ⅜ inch wide and 1½ inches high, spaced 2 inches apart. The riffles are built in the form of trays 3 feet long, so that they can be removed readily during clean-ups. By means of an eccentric drive, the sluice boxes are driven in tandem, 120 strokes per minute. A considerable amount of black sand is present in the gravel, which packs in the riffles, and necessitates frequent

clean-ups. The head end of each sluice is cleaned about every
two hours and a general clean-up is made every 16 hours. The
concentrates obtained from the boxes are cleaned by hand panning with mercury.

Water for the washing plant is pumped from a reservoir
excavated in the bed of Lynn Creek above the placer ground.
Operations are handicapped by lack of water during most of
the year. In 1935, the surface water was available for about
three months in the spring of the year. It is estimated that the

Figure 25. Loosening gravel with spring-tooth harrow, Lynn District.

amount of water required to wash six yards per hour is 60
gallons per minute, or 600 gallons per cubic yard of gravel.

During the operating season of 1935, approximately 10,000
cubic yards of alluvium were handled by the shovel. Two-thirds
of this amount consisted of overburden, which was too low grade
to be treated at a profit. The overburden is stripped to a depth
varying from 2 to 7 feet, average 5 feet, and the pay streak
varies from 1½ to 4 feet in thickness, average about 2½ feet.
The average grade of the gravel washed is about 40 cents per
cubic yard. With this grade of gravel and with the above conditions it is estimated that a profit can be made in stripping two
yards of overburden to each yard of gravel washed. The pay
streak is under 12 feet of overburden in places, but this amount
of stripping cannot be done at a profit.

The overburden is mainly soil with some water-worn boulders.

The pay streak is a mixture of wash gravel and sand lying on a false bedrock of silt. In places the gravel is loosely cemented. True bedrock has never been reached. In former years a shaft was sunk on the property to a depth of 75 feet and, after passing through the pay streak near the surface, this shaft was entirely in barren silt. The channel varies from 18 to 33 feet in width. The gold is all fine, no piece being larger than a pinhead. Fineness of the gold averages 936.

The Arrowhead placer is in Sheep Creek Canyon on the west

Figure 26. Power-driven sluice box, Lynn District.

slope of the mountains. This property has been worked for a number of years by K. C. Nelson. Production has been consistent and ranges from 150 to 250 ounces per year. Both dry-washing and wet-washing methods are employed. During the season when water is available, an unusual method is employed by Nelson for recovering the gold. The equipment consists of a reservoir for impounding water, a harrow, and a power-driven sluice box. The gravel in the creek bed is loosened with a horse-drawn spring-tooth harrow while water is flowing over it. (See figure 25.) In this way a large proportion of the soil and clay is washed from the gravel. The partly washed gravel is then moved into a pile by a horse-drawn scraper. The gravel is shoveled from the pile by hand into the power-driven sluice box 16 feet long, 10 inches deep, and 12 inches wide. The box is set on rockers and shaken from side to side by means of an eccentric rod connection driven by a 1½-horsepower gasoline engine. Its capacity is about four cubic yards per hour. (See figure 26.) Hungarian riffles made of wood strips, 1 by 2 inches, are placed 2 inches apart in the lower part of the sluice. In the upper end the riffles consist of 2-inch plank with numerous holes 1 inch in diameter. This board is supported two inches above the bottom of the box by a wooden grating. This short section of riffles is effective in saving the gold, and about two-thirds of the gold recovered is caught at the upper end of the sluice box. The upper end of the sluice is cleaned up more frequently than the lower end. The lower riffles are made in two 5-foot sections. The amount of water required is small as compared with other gravel-washing equipment, but, notwithstanding the low water

Figure 27. Power-operated rocker, Lynn District.

consumption, the tailings from the box are said to contain less than five cents per cubic yard.

Early in the season when more water is available, ordinary sluice boxes are used. When the supply of surface water ceases, usually the latter part of June, dry-washing machines are resorted to and the work is continued until fall. The placer season lasts six to seven months per year. The tailings from dry-washing machines are reworked by sluicing when water is available. The fineness of the gold averages about 930.

George Graves has worked adjoining ground with a patented power rocker for recovering the gold. (See figure 27.) The outside dimensions of the rocker are 27 inches wide by 49 inches long at the top and 21 inches by 49 inches at the bottom. It is 24 inches high in front and 21 inches in the rear. This machine rests on wooden rockers. The screen at the head end is iron plate perforated with round holes $5/8$ inch in diameter, staggered on $2\frac{1}{2}$ - inch centers. Beneath the screen are three aprons made of wood and canvas placed at angles so that the fine gravel that falls through the screens travels over each apron, following a zig-zag course from end to end of the box until it reaches the outlet at the bottom. Each apron has a fall of three inches per foot. The frame of the aprons is made of $1\frac{3}{4}$ x $\frac{3}{4}$-inch wood strips to support the canvas. On top of the canvas are $\frac{1}{2}$ x $\frac{1}{4}$ -inch wood riffles. The box is rocked by an eccentric arm at the rate of 40 strokes per minute. The length of the stroke is six inches. Nearly all the gold is caught on the first apron of the rocker. The average capacity of the machine, run by two men, is one cubic yard per hour; but when the gravel is free from clay the capacity of the machine increases up to as much as three cubic yards per hour. The cost of the rocker, including the power equipment, is about $160.

Bulldog placer is at the head of Lynn Creek on the site of the original discovery of placer gold in the District. This placer has been a fairly consistent producer for a number of years. In 1935, the placer was worked on a leasing basis. The Seymour lease, employing four men, has worked the ground for several seasons with good results. A power-driven sluice box is employed to recover the gold.

The power sluice box is a side-shake affair similar in construction to the one shown in figure 26. The box is 12 feet long, 12 inches high, and 12 inches wide and is made of 2-inch plank. The inside of the box is lined with 28-gage galvanized iron. Riffles are wood strips 1 by $1\frac{1}{2}$ inches in section, placed longitudinally in the box. The tops of the riffles are armored with strap iron to prevent wear. Two 4 - foot sections of riffles are used with $\frac{1}{2}$ foot of space between the head of the box and the first section and the same amount of space between the two sections. The box is rocked on a shaft 8 feet long and $1\frac{1}{2}$ inches in diameter. A bearing is used at each end of the shaft. The box is rocked by an arm connected to the crankshaft of an old Fairbanks - Morse engine. The crankshaft is belt driven by a $1\frac{1}{2}$-hp. Fairbanks-Morse gasoline engine at a speed of about 80 strokes per minute. Some hard clay containing gold is present in the gravels, and to assist in the disintegration of the clay several 16-pound steel balls are placed in the box. The cost of this power sluice box, complete, is as follows:

Engine (new)	$74.50
Bearings	5.00
Shaft	2.00
Lumber	7.00
Lining	10.80
Labor	15.00
Total	$114.30

With four men working, the rocker has treated an average of eight cubic yards per day. With gravels that are less consolidated, this capacity can be increased considerably. Water consumption is estimated to be about 100 gallons per cubic yard of gravel sluiced. In 1935, the Seymour lease had a two-month run with surface water, and for several additional months water was piped from several small springs at the head of the canyon. The gold is concentrated largely in the four feet of gravel above bedrock. The depth to bedrock averages about 15 feet. The overburden is removed with horse team and Fresno scraper. The overburden carries some gold which it is reported could be sluiced at a profit, but it is too low grade to be worked with power sluice box. The grade of the material worked ranges from $1.50 to $8 per cubic yard. In some places as high as $2\frac{3}{4}$ ounces of gold was recovered in a day's run, but such rich spots are unusual.

Several other small placer operations were carried on in Lynn Creek Canyon during 1935. The canyon is about four miles long and the main channel, which averages about 25 feet wide, contains very little virgin ground, as most of the placer gravels were worked in former years by sluicing or rockers during the spring of the year and hand dry washers during the summer months. The placer ground is held by a number of owners, who hold from one to four claims each.

In general, the gravels in Lynn Creek Canyon range from 10 to 28 feet in depth. A large number of medium and large sized boulders occur, with some clay. The largest boulders weigh over a ton. The gold shows signs of wear and, as a general rule, it is fairly coarse, although no large nuggets are found. The largest nugget found in the District some years ago was worth $21. In 1935 the largest nugget found had a value of $13. In fineness the gold ranges from 928 to 934, and some black sand is associated with it.

The side hills adjacent Lynn Creek Canyon offer opportunities for prospecting, as gold can be panned from the detrital material. The south side of the canyon appears to carry better values than the north side. In 1934, Lee Tolin, working with one man on the south side of the canyon, about half a mile from the head, recovered $3,000 by dry washing. The gravel worked by Tolin ranged from three to five feet in depth, and the pay was confined mainly to one and one-half feet of alluvium directly above bedrock. Bedrock is rough. The gravels on the side hills contain a fewer number of boulders and less clay than those in the bed of the creek. In addition, the material is fairly dry so that a good recovery can be made by dry washing methods.

MAGGIE CREEK

The Maggie Creek District, also known as the Schroeder or Susie Creek District, is in the Tuscarora Range in northwest Eureka County near the west boundary line with Elko County, about 10 miles northwest of Carlin. A little placer mining has been done from time to time at the head of Maggie Creek, but no large production of placer gold has been recorded.

HUMBOLDT COUNTY
GOLD RUN DISTRICT

The Gold Run or Adelaide District is in southwestern Humboldt County on the east slope of the Sonoma Range. The District was organized in 1866. The amount of placer gold produced has been small. It is reported that in the early days some hydraulicking was done in the ravine below the Adelaide mine.

The Bonanza, Gold Run, and several other small properties about ten miles south of Golconda were worked by lessees in a sporadic manner from 1914 to 1922. The gravel was mined by sinking shafts and drifting along bedrock. Rockers were employed for recovering the gold. In the last three years there has been a little activity in the District, and rich gravel was being worked on one claim.

On the east slope of the Sonoma Range, about nine miles south of Golconda, the Ontario-Nevada Mines, Inc., operated a placer for a short period in 1935. The placer plant consists of a Bucyrus - Erie power shovel, type GA – 2, trucks for hauling material to the washing plant, and a washing plant.

The washing plant comprised a trommel 16 feet long and 4 feet in diameter, the first 12 feet being equipped with a punched screen having holes ⅝ inch in diameter, and the other 4 feet with 1-inch holes. Below the trommel is a washing cylinder about 10 feet long and 3 feet in diameter for disintegrating the clayey material in the minus ⅝-inch product from the trommel.

After the material is washed it is diverted to the riffle deck, made of three parallel boxes each 12 feet long and 3 feet wide equipped with Hungarian riffles. The plant is shown in figure 28.

The tailings from the washing plant pass by gravity to the boot of a belt - and - bucket elevator that discharges into a bin. From the bin, the tailings are hauled away by truck.

The alluvium is mined with the power shovel, loaded into a truck, and hauled a few hundred yards to the washing plant. An inclined ramp permits dumping the trucks on the grizzly over the trommel feed bin. The grizzly is made of 20 - pound rails spaced with 3-inch openings. The oversize passes into a second bin, from whence it is hauled away by trucks.

Water for the washing plant is obtained from a well sunk

Figure 28. Washing plant at property of Ontario-Nevada Mines, Inc., Gold Run District.

above the placer ground. It is pumped into a wood - stave storage tank 20 feet in diameter and 5 feet high. Water flows from the tank to washing plant by gravity.

The alluvium is mainly soil and sand with a few small, angular rock fragments, and from 10 to 14 feet deep. Bedrock is schist. At the time of visit, in the latter part of November, 1935, the property was closed down.

PARADISE VALLEY DISTRICT

The Paradise Valley District is on the east slope of the Santa Rosa Range, eight miles northwest of the town of Paradise Valley. It was also known as the Mount Rose District at one time.

The lode and placer deposits were discovered in 1868. The bulk of the production has come from lode deposits. Small productions of placer gold were made from 1909 to 1915, but in recent years very little placer mining has been done in the District.

REBEL CREEK DISTRICT

The Rebel Creek District, also known as the New Goldfields or Willow Creek District, lies on the west slope of the Santa Rosa Range, about 54 miles north of Winnemucca, Nevada. The placer and vein deposits in the District have been prospected intermittently since the seventies, but no discovery of major importance has ever been made. Placer gravels have been worked on a small scale on the west slope of the Santa Rosa Range from Pole Creek on the north to Rebel Creek on the south, a distance of 17 miles. On Pole Creek, in the early days, considerable quartz float carrying gold was found and a 5-stamp mill was erected to treat it. It is reported that $30,000 worth of gold was recovered from this float. No vein was found.

Placer gravels have been found and worked in Willow Creek and Canyon Creek, the latter being due west of Spring Peak. The Willow Creek placers were worked in the early days by Chinese. Several years ago a sluicing plant operated on this creek for several seasons when water was available, but the results of this work were not encouraging. Some desultory dry washing was done in the same vicinity in 1935, but gold was not found in paying quantities.

The placer deposits in the District presumably were derived from the pre-Tertiary quartz veins in the slates and other rocks of the Santa Rosa Range.

KING'S RIVER DISTRICT

The King's River District is an unorganized District 45 miles northwest of Orovada, Nevada, near the Oregon - Nevada boundary line. In the early days, probably about the late seventies, the Chinese are reported to have worked the placer gravels along Chinese and Horse Creeks, which are tributary to King's River. In 1935, a small sluicing operation was carried on along Chinese Creek by two men. Gravel was shoveled into sluice boxes by hand. This operation was handicapped by inadequate water.

VARYVILLE DISTRICT

The Varyville District, sometimes called the Columbia or Leonard Creek District, is in northwest Humboldt County, at the south end of the Pine Forest Mountains, about 100 miles northwest of Winnemucca. It was originally organized as the

Columbia District in 1875, and lode mining was actively conducted for a number of years. The placers were discovered subsequent to the lodes and have been worked intermittently on a small scale up to the present time.

In 1931, 1,500 acres above the Montero Ranch along Leonard Creek for a distance of two miles was under investigation as placer ground by a San Francisco group of men, who organized a company called the Leonard Creek Placers, Ltd. The ground was tested, and sampling indicated that a considerable yardage is available having an average gold content of 50 cents per cubic yard. The average depth of the gravel is about 20 feet. Successful exploitation of the placer ground may depend upon obtaining a supply of water from the Leonard Creek Ranch, which owns the water rights on Leonard Creek. The Leonard Creek Placers, Ltd., sank several wells for water at the lower end of Leonard Creek Canyon, but data on the results of this work are not available. The company has never installed equipment for working the placers on an extensive scale.

In 1932 a number of miners were sluicing, rocking, or dry washing on Teepee and Snow Creeks, tributaries of Leonard Creek. During the spring, water for sluicing is obtained from melting snows. Later in the year, when this water supply fails, dry-washing methods are employed. · In 1935 there was very little placer activity in the District.

DUTCH FLAT DISTRICT

The Dutch Flat placer is 18 miles northeast of Winnemucca and 18 miles north of Golconda, on the west slope of the Hot Springs Range. Placer gold was discovered in the District in 1893 by Fred G. Wendel. According to Mr. Wendel, the placer gold produced in the District has been worth about $200,000, nearly all of which was recovered by working with rockers. For several years following the discovery of placer gold in the District, about 20 men were employed in working the gravels with rockers with good results. In the first year, production is said to have been $75,000. Water for the rockers was hauled in wagons from Spring Canyon, about one mile from the placer ground. The water was obtained from a well sunk to a depth of 25 feet.

In 1904, a company organized in Salt Lake City, called the Dutch Flat Gold Mining Company, attempted to work the placer on a more extensive scale than was possible by hand methods. Twenty acres sampled by the company is reported to have averaged 31 cents per cubic yard, and contained 375,000 yards of gravel, averaging 15 feet in depth. The equipment installed consisted of a trommel 16 feet long and 4 feet in diameter, sluice, and tailings elevator. The gravel was mined and transported to the machine by dragline scraper. The oversize from the screen was discharged to the waste pile by the tailings elevator. The undersize was treated in the sluice. This venture was not profitable as the company could not obtain an adequate water supply for the washing plant.

In 1909 and 1910 the placer ground was leased by Wendel and associates to Chinese placer miners. About 20 Chinese were employed in working the placer, and numerous shafts and tailings piles attest to the diligence with which this work was conducted. (See figure 29.)

Figure 29. Chinese placer diggings, 1909 and 1910, Dutch Flat placer, Humboldt County, Nevada.

In later years the placer has been worked in a small way either by Wendel or lessees. Wendel is the sole owner of 11 placer claims in the District, which cover the best ground. From his placer holdings Wendel has made a living for about 40 years.

The gravel is drift mined and the material taken out above bedrock is washed in rockers. A small flow of water, amounting to about one gallon per minute, is obtained from a tunnel at the head of the ravine where placer gold is found. It is brought to the placer ground by gravity through a 2-inch pipe.

The placer is about one and one-half miles long and varies in width from 300 to 2,000 feet. There is no regular channel wherein the gold is concentrated, and good values have been found on the hillsides adjacent to the ravine as well as in the ravine itself. Over 100 shafts have been sunk on the property. These shafts vary in depth from 6 to 22 feet, averaging about 12 feet to bedrock. One shaft sunk near the mouth of the ravine in placer material has a depth of 35 feet. The bedrock is composed of schist, rhyolite, and granite. The gravels are a mixture of detrital material consisting of angular rock fragments and sand, which in places is cemented with clay. The largest rock fragments are less than six inches in diameter. The greatest concentration of gold is on bedrock, the pay streaks varying from six inches to three and one-half feet thick. Both fine and coarse gold occur. The coarse gold is rough and angular and some of it is attached to a quartz matrix, indicating that it has not traveled far. The largest nugget ever found in the placer had a value of $180. A large amount of black sand and some cinnabar is associated with the gold. The gold has an average fineness of 940. According to Wendel, in places the placer material will average about $1 per cubic yard. The placers have been derived from the erosion of the quartz veins in the vicinity.

Economical exploitation of the placer depends in great measure on the development of an adequate water supply. According to Wendel, a flow of water amounting to about three miner's inches can be developed in Mill Canyon, three miles from the placer ground. This water could be brought to the property by gravity flow.

LANDER COUNTY
BATTLE MOUNTAIN DISTRICT

The Battle Mountain District is in the Battle Mountain Range in north-central Lander County. When originally organized, the District was in Humboldt County, but was ceded to Lander County in 1873. It covers a wide area and includes the placers in Copper and Rocky Canyons, Cottonwood Creek, Willow Creek, Copper Basin, Galena on Duck Creek, and the vicinity of the old abandoned mining camp of Bannock. Lode deposits in the District were discovered in 1864, but the placers were not worked until about 1909.

Although lode mining has been the principal industry in the District, a large number of small placer properties have been productive from 1909 up to the present time. From 1910 to 1935, inclusive, the placer production has been approximately $900,000. In 1910, the year of the first recorded placer production, the output was $5,627, and it increased steadily until it reached a maximum of $160,596 in 1915. From 1915 it declined slowly to $4,406 in 1929. In recent years the District has witnessed a revival in placer mining, and a large number of small-scale operations have been active, generally with fair returns. In 1935, about 50 men were engaged in placer mining in the District.

Dry washers and hand rockers have been employed generally for small-scale operations. In some instances, water has been hauled as far as five miles to work the gravels in rockers. Where the gravels are deep and the values are concentrated largely on bedrock, drift mining is done. During the spring of 1912 and 1913 small-scale hydraulic operations were carried on in Copper Basin by Andrew Kinneberg and Jasper Vail of Battle Mountain. The Dahl and Christensen placers on the alluvial fan at the mouth of Copper Canyon have been worked with power equipment in recent years.

The most productive areas are at the mouth of Copper Canyon and in Black Canyon, near Bannock. Much of the gold is coarse and ranges from 830 to 890 in fineness. The largest nugget found in the Bannock area weighed 3.51 ounces. In 1931 a nugget was found in Copper Canyon that had a value of $150.

Most of the placer gulches in the District are V-shaped, and the gravels are subangular rocks and sand roughly stratified by intermittent stream action. The richest gravels are found usually on or directly above bedrock, but in some cases values are found throughout the gravels up to the surface. Limonite, quartz, and pyrite are occasionally attached to the gold, indicating that the gold is local in origin and has not traveled far from its source. At Copper Canyon boulders of copper ore have been found in the placer channels, and it is believed generally that the gold in these deposits was derived from the copper - gold deposits a short distance east of the placers.

The principal producers in the Battle Mountain District are the Dahl and Christensen placer and the Homestake Consolidated Mining Company, the latter owned by the Copper Canyon Mining Company. Both properties are at Copper Canyon. Placer gold was discovered at the mouth of Copper Canyon in 1913 by James Dahl, a former placer miner in the Klondyke rush, working in partnership with H. C. Christensen of Battle Mountain. A short time after this discovery as many as 90 men were employed by Dahl and Christensen to drift mine the placer channels. The gold was recovered in hand rockers. With the exhaustion of the richer gravels, the placers were taken over

4

by Good, Kinneberg, and Wilson, who constructed a dragline
and sluice for working the placers more economically than was
possible by hand methods.   Water is scarce in the proximity of
these placers, which has proved a drawback ever since work on
the placers was begun.   Water for sluicing was obtained from
Willow Creek.

About 1930 the Dahl and Christensen placer was acquired by
Dudley Wilson, one of the three former owners.   The Wilson
placer operation was carried on by sluicing for several seasons.
The ground was mined by drifting on bedrock.   The mine was
entered by tunnel parallel to the course of the canyon.   This
tunnel is 2,300 feet long, 1,600 feet of which was driven in
gravel before the channels were encountered.   Three separate

Figure 30.   Gravel bins above sluice at Wilson placer, Battle Mountain District.

channels were worked by lateral drifts from the main tunnel.
The gravel was trammed to the surface and dumped into a
series of bins from an elevated trestle a short distance from
the portal of the tunnel.   The trestle and bins are shown in
figure 30.   From the bins, the gravel was fed into a sluice several hundred feet long and equipped with light steel T-rail riffles.
The sluice had a grade of 7 inches per 10 feet and was 16
inches wide and 12 inches deep.   Water was supplied to the sluice
from a reservoir of about 200,000 gallons capacity, to which the
water was piped from Willow Creek, a distance of three miles.

In addition to the foregoing operation, old placer tailings were
reworked by sluicing.   The tailings were transported to the sluice
with a Sauerman dragline scraper having a capacity of one-half

cubic yard. The average distance of scraper haul was 100 feet, the maximum distance 200 feet. The scraper was operated by a 22 hp. gasoline engine mounted on a truck. The gravel composing the tailings was subangular, and the largest boulders did not exceed six to eight inches in diameter. The average depth of the tailings worked was six feet, and although the gold content was low the operation was profitable. Figure 31 shows the scraper discharging into the sluice.

In 1935 the Dahl and Christensen placer was worked by the Grand Hills Mining Company, organized by a group of men from Oklahoma City, Oklahoma. The placer property comprises 26 claims of 20 acres each, unpatented. The company employs an average of 15 men. The placer channels are mined with

Figure 31. General view of Wilson placer operations, Battle Mountain District.

power shovels, and the pay gravel is treated in a mechanical washing plant erected by the company in February, 1933. Due to lack of water, the plant was idle for one year.

The average depth of the gravel in the placer channels is about 50 feet. In mining the gravel, the upper 35 feet is removed with a 1¼-cubic-yard capacity Bucyrus-Erie gas-air shovel and Chevrolet trucks holding about two yards each. The overburden is hauled about 200 feet and the cost of stripping is reported to be 8 cents per cubic yard. Although the top 15 feet of the overburden averages 5 cents per cubic yard and the next 20 feet 20 cents per cubic yard, it is too low grade to be worked. The 15 feet of gravel directly above bedrock averages

$1.50 per cubic yard. The pay gravel is mined with a Lorain shovel of 1¼-cubic-yard capacity driven by a gasoline engine, and is loaded into Chevrolet dump trucks, which transport it to the washing plant about 3/10 mile distant. The stripping and mining operations are shown in figure 32.

The washing plant has a capacity of 45 cubic yards of gravel per hour and is run on a one-shift-per-day basis. A general view of the washing plant is shown in figure 33. The trucks dump on an inclined grizzly, which is made of 30-pound rails spaced with 6-inch openings. The oversize is discharged into a dump truck, which hauls it several hundred feet to the waste dump. The undersize passes into a feed bin having a capacity

Figure 32. Stripping and mining placer gravel at Grand Hills Mining Co. property, Battle Mountain District.

of 20 cubic yards. At the bottom of the bin a home-made shaker feeder with 4-inch stroke feeds the gravel onto a 3-ply inclined conveyer belt 28 inches wide, which delievers it to the trommel.

The trommel is 22 feet long, 6 feet in diameter and is equipped with four sections of screen each 4 feet in length. The first three sections have ⅜-inch diameter holes, and the fourth section has ½ x ¼-inch holes. In the center of the trommel, several feet apart, there are two sheet-iron rings 6 inches high to retard the flow of gravel. To break up any gravel that may be loosely cemented, three equally spaced 3½-inch angle irons are riveted longitudinally inside the trommel. The trommel water spray consists of a 3-inch diameter pipe closed at one end and perforated by a number of 3/16-inch diameter holes. The trommel slopes 1½ inches per foot and it revolves on flanged rollers at a speed of 9 revolutions per minute.

The trommel oversize is discharged by gravity into a bin that holds 8 cubic yards. From the bin the material is loaded into dump trucks and hauled to the waste dump. In the original design of the plant a tailings stacker was used, but, due to freezing weather and the comparatively level site on which the plant is built, the tailings stacker proved unsatisfactory and was superseded by truck haulage.

The undersize from the trommel is split into four streams, each stream passing into an Ainlay centrifugal gold separator. Each of the four Ainlay bowls is 36 inches in diameter and

Figure 33. General view of washing plant of Grand Hills Mining Co., Battle Mountain District.

revolves at a speed of 91 revolutions per minute. The principle of the Ainlay bowl is simple. The water and fine gravel pass through a pipe to the bottom of the bowl, and centrifugal force causes the gold and black sand to adhere to the sides while the lighter material passes upward and over the rim. To prevent the gold particles from passing upward and over the rim, the bowl is equipped with peripheral riffles. Ninety percent of the gold recovered is caught in the bowls and the remaining 10 percent is trapped in a sluice box 40 feet in length, which carries away the Ainlay tailings. The bowls are cleaned up once or twice per shift, depending on the richness of the material treated. About five large gold pans of concentrates are recovered from the bowls each shift. This material is cleaned by panning with mercury and the gold is recovered by retorting in a small retort.

The sluice box below the Ainlay bowls is equipped with metal lath riffles. The final tailings have been sampled and assayed several times and the amount of gold lost in the tailings is very small.

The trommel, conveyer belt, and Ainlay bowls are driven by a 15-hp. International Harvester tractor engine. A 10-hp. Rex engine is used to drive a 4-inch centrifugal pump, which lifts the water about 18 feet to the trommel. The plant requires about 400 gallons of water per minute, or about 535 gallons per cubic yard of material treated. Water is obtained from Willow Creek Springs through five miles of open ditch. In the fall of 1935 the company was drilling several wells on a flat two and one-half miles from the placer plant in an attempt to increase the water supply. If sufficient water can be developed, the capacity of the plant will be increased. With the present layout the operating cost is about 50 cents per cubic yard of gravel treated.

The placer ground is tested in advance of mining with drill holes spaced on 35 - foot centers. An Armstrong drill with a 6-inch cutting shoe is employed for the work. Drilling is done on a two-shift basis. Drill cuttings are sampled with a rocker.

The placer gravel consists largely of subangular rock fragments and sand, though a few large boulders occur, some weighing as much as five tons. Bedrock is a smooth sedimentary formation. Gold, both fine and coarse, occurs with some black sand. The fineness of the gold ranges from 875 to 890. The largest nugget found in 1935 had a value of $4.

About 18 men were placering in 1935 in Copper Canyon, above the ground owned by the Grand Hills Mining Company. Gold was recovered with hand rockers, dry washers, and one power sluice box. Dry washing was employed in reworking old dumps.

The main placer channel in Copper Canyon was worked pretty thoroughly by drift mining in former years. The canyon is about one and one-half miles in length and from 125 to 300 feet in width. Bedrock, which is a decomposed porphyry, is from 30 to 60 feet below the surface. The gold is concentrated chiefly in the two to six feet of gravel directly above bedrock. Medium and large sized boulders are numerous. The gravel is cemented and in drift mining, blasting is necessary to loosen it. Most of the gold is coarse and fairly well worn, its fineness averaging about 830.

In 1935 George Bowen and two partners were leasing on ground owned by the Copper Canyon Mining Company, paying a royalty of 12 percent of the value of the gold recovered. Operations were confined to a small section of virgin ground on a bench. The shaft was down 45 feet to bedrock and an average width of eight feet of gravel was drift mined. An 8 - hp. Fairbanks-Morse geared hoist was used. Holes were drilled in the gravel by means of a pointed bar and sledge hammer. Forty percent gelatine dynamite was used for blasting. With three men working, about 5½ yards of gravel was mined per half shift; the gravel was washed on the other half shift.

The tripod headframe over the shaft and the power sluice box used to recover the gold are shown in figure 34. The power sluice box was driven by a 2 - hp. John Deer gasoline engine. Water consumption was estimated to be about 100 gallons per minute. Water was obtained from the domestic water supply

Figure 34. Power-operated sluice box, Bowen lease, Copper Canyon, Battle Mountain District.

of the Copper Canyon Mining Company. Hungarian type riffles were used in the sluice box. According to Bowen, the gravel must average $4 per cubic yard in order to make wages, but the Bowen lease netted the operators better than this. The clean-up for a one-day run at the time of the writer's visit was 1 ounce, 8 pwt. of gold. The largest nuggets found in 1935 weighed 13 pwt., 7 gr., and 12 pwt., 18 gr.

STEINER CANYON DISTRICT

The Steiner Canyon or Bobtown placer is about midway between Austin and Battle Mountain near Bobtown on the Nevada Central narrow-gage railroad. In the early seventies gold was discovered here in a well dug to supply water for a stage station. At the point in the well where placer gold was discovered an abundant flow of water also was encountered, but nothing further was done until recently, when an attempt was made to sink the well to bedrock. The flow of water proved too great to be handled with the available equipment, and the prospectors attempted to sink a shaft on a site farther up the canyon. Here, also, operations were impeded by a large flow of water encountered at a depth of 40 feet. Placer operations at Steiner Canyon are still in the prospecting stage and no placer gold has been produced.

TENABO DISTRICT

The Tenabo District, also known as the Campbell, Bullion, and Lander District, is in the northern part of Lander County about 25 miles miles southeast of Battle Mountain, Nevada. Placer gold has been found here over an area of 15 square miles, but

Figure 35.   Portable dry-washing machine of special design used at Camp Raleigh, Tenabo District.

most of the prospecting and development work has been confined to several gulches near Camp Raleigh. The best ground is in Triplett and Mill Gulches near Camp Raleigh, which is on the west side of Crescent Valley, 22 miles south of Beowawe, Nevada. Placer gold was discovered in Mill Gulch in 1916 by A. E. Raleigh, for whom the camp was named. The Mill Gulch placer was worked for several years after its discovery by Raleigh, who used a rocker to recover the gold. Water was hauled by burro from Indian Springs, two and one-half miles north of the placer ground. Since 1916, the District has produced small amounts of placer gold annually. Accurate figures on the total value of placer gold produced are not available, but according to information received from men who have been active in the District for a number of years, the amount is probably in the neighborhood of $25,000. Most of this gold was produced in the last five years.

Placer mining has been carried on with dry washers, rockers, and several machines of special design. In 1930 a Giffen placer machine was operated for a time in Triplett Gulch with some success. This machine is designed for use in localities where water is scarce. It is reported that the water consumption in a test run was about 80 gallons per cubic yard of gravel treated. The machine consists essentially of either a shaking or a stationary hopper fed by an elevator. Water under pressure is fed into the hopper with the gravel, and from there the material is

Figure 36. Two-unit dry-washing machine, power driven, used at Camp Raleigh, Tenabo District.

fed to a shaking table having a side motion and set at a pitch of three inches per foot. The deck of the table is equipped with Giffen riffles. A trommel may be used to screen out the larger size boulders before washing.

In 1934 the Earl Corporation of Minneapolis tested a drywashing machine of unique design. This machine, shown in figure 35, ran only a few days. The company owns two claims in the District.

In 1934 and 1935 Herman Rieck, William Bown, and Carl Adeen worked placer ground in Triplett Gulch under a three years' lease on a royalty basis of 10 percent of the gross bullion output. The ground consists of nine claims of 20 acres each owned by Mrs. George W. Triplett of Battle Mountain. In 1934 250 ounces of gold were recovered and in 1935, due to adverse weather conditions, about half this amount was produced. A power dry washer with a capacity of 50 cubic yards per eight hours is used to recover the gold. The dry washing plant is shown in figure 36.

On the Bown lease the placer material is mostly sand and soil containing about 5 percent of angular rock fragments not exceeding 5 inches in diameter. The depth varies from 1 to 12 feet, and averages about 6 feet. Bedrock is fairly smooth andesite and diorite. The gold is both fine and coarse, with an average fineness of 910. The operators state that the average value of the placer material treated is $2 per cubic yard from the surface to bedrock. In places as high as $30 per cubic yard has been recovered.

In working the placer, the ground is first broken to a depth of about one foot with a scarifier pulled by a 15-hp. caterpillar tractor. It is then worked over with a spring-tooth harrow and tractor. An interesting feature of the harrowing operation is the method of breaking loosely cemented lumps of sand and soil. For this purpose a flat-bottomed stoneboat made of 2-inch plank is loaded with stone and dragged behind the harrow. This serves to break the small lumps of clayey material that may contain gold that would otherwise be lost when the material is run through the dry washer. After the material is harrowed and has had an opportunity to dry thoroughly in the sun, it is hauled to the grizzly ramp at the dry washer by a 19-cubic-foot capacity Master Rotary scraper and tractor. The grizzly is made of 20-pound rails spaced with 3-inch openings. The undersize falls into the loading hopper of a Jeffrey portable conveyer, which elevates the material 12 feet to the hopper above the battery of two dry-wash machines. The conveyer is driven by a 3-hp. Fairbanks-Morse gasoline engine. Above the hopper is an inclined woven-wire screen of $\frac{1}{2}$-inch mesh. The screen oversize falls to the ground in front of the dry washer and is hauled away with the dry-wash tailings by tractor-operated scraper.

The dry-washing machine consists of two air jigs operated by bellows. The jigs are similar in design to those in general use throughout the arid regions of the southwest. They are driven by belt drive from a 2-hp. Fairbanks-Morse gasoline engine (not shown in figure 36) mounted 50 feet in the rear of the machine in order to avoid the dust. An idler pulley is used in the center of the belt. Only three men are required to handle 50 yards of placer material in eight hours with the equipment described.

Operations are handicapped, however, by weather conditions so that the placer season is rather short. A light rain is sufficient to stop the work until the ground dries. In the fall of 1935 the operators were using a power-driven dry washer (shown in figure 37) to treat selected portions of their ground. This material is mined near bedrock and is comparatively dry, so that a good saving can be made.

About half a mile below the Bown lease a company, called the Desert Placers, Inc., incorporated under the laws of Nevada, is working placer ground consisting of three claims and a fraction leased from Bown, et al. The Desert Placers, Inc., is controlled by J. Goulden Evans and A. H. Kelson of Salt Lake City. A view of the placer plant at this property is shown in figure 38. The

Figure 37. Power dry washer used at Camp Raleigh, Tenabo District.

plant has a capacity of 100 cubic yards per shift and requires water for concentrating. This plant is the outgrowth of considerable experimental work so that continuous operation for a long period has not been possible. In 1934 and 1935 about 8,000 cubic yards of placer material were treated.

The placer material is composed chiefly of sand and gravel with a small amount of angular rock fragments. At least 75 percent of the alluvium will pass through a $3/8$-inch mesh screen. The depth to bedrock varies from 1 to 7 feet, averaging 5 feet. Bedrock is schist and andesite. Part of the ground has been sampled by shafts and open cuts, and the average value, as determined by panning, is 60 cents per cubic yard. The gold is fine and angular, with a few small nuggets. The two largest nuggets found had a value of $5 and $3. Some of the gold is so fine that it will float on water. A small amount of black sand occurs.

In working the placer the ground is broken up with a 5-toothed scarifier drawn by a 30-hp. caterpillar tractor. A scraper of 42 cubic feet capacity is used to haul the material to the washing plant. The scraper is drawn up an earth ramp and unloaded over a grizzly made of 30 - pound rails with 4 - inch openings. Below the grizzly is a hopper that holds 7 cubic yards. From the hopper the material is fed onto a 5-ply conveyer belt 24 inches wide and 40 feet long, which discharges into a 4-cubic yard bin above the trommel. The alluvium is fed into the trommel by a belt feeder 3 feet between pulley centers.

The trommel is 12 feet long, 4 feet in diameter, and is divided

Figure 38.   Washing plant operated by Desert Placer, Inc., Camp Raleigh, Tenabo District.

in two sections; the first or disintegrating section is of plain sheet iron and the second section is covered with ⅜-inch mesh woven-wire screen. The plus ⅜-inch product is conveyed to a waste pile by a 3-ply Jeffrey stacker belt 14 inches wide and 36 feet long. The undersize passes by gravity to a shaking table with side motion placed beneath the trommel. The table is 10 feet long and 3½ feet wide. The deck of the table is covered with gold-saving corduroy, on top of which are strapiron riffles ¼ inch wide and 1½ inches high, spaced 1 inch apart.

The table tailings go to a small sump, which is connected to a 4-inch Wilfley sand pump. Tailings are pumped to a series of settling ponds, and after settling some of the water is reclaimed. The reclaimed water, which contains about 10 percent solids, is returned to the trommel by a 2-inch centrifugal pump. Water is fed into the trommel through a 2-inch pipe into which a number of small spray holes have been drilled.

A clean-up is made each shift. The concentrates are cleaned by panning. About 95 percent of the gold is recovered by panning without using mercury. A small amount of fine gold in the concentrates is recovered by amalgamation and retorted over an open fire.

There is some loss of fine gold in the tailings, as shown by careful panning. A 50-pound sample of the sand and slime portion of the tailings was taken by the writer, and a fire assay gave 0.049 ounce of gold. At the present price of gold, this would amount to $1.71 per ton. This sample indicates that there is considerable loss in the tailings. It may be that some of the gold is encrusted with quartz so that it cannot be recovered by gravity methods. However, a single sample is not conclusive, and additional samples should be taken to determine the amount of gold lost in the tailings.

The placer plant is operated by a 30-hp., 440-volt, 3-phase, 60-cycle Fairbanks-Morse Diesel engine-generator set. Power distribution is as follows:

	HORSEPOWER
Tailings pump	10
Shaking table	3
Trommel	3
Tailings stacker	3
Main belt conveyer	5
Water return pump	3
Belt feeder	1

In addition, a 2-kw. transformer is used to reduce the voltage to 110 for the lighting circuit. Six 300-watt flood lamps are used for night work.

Cost of fuel oil laid down at plant is 9 cents per gallon.

The placer plant consumes about 60 gallons of water per minute. The water is pumped from a well sunk in the valley below the placer ground, and is delivered to the plant through 11,000 feet of 4-inch iron pipe by multistage turbine pump driven by a 20-hp. Waukesha gasoline engine. Total static and friction head is 460 feet.

Five men are required to operate the plant on a two-shift basis, and it costs 30 cents per cubic yard to treat the alluvium.

The Mill Gulch placer, owned by A. E. Raleigh, and comprising five 20-acre claims, is half a mile north of Triplett Gulch.

Production of placer gold from the Mill Gulch placer, by small-scale operations, is reported to be about $2,000.

Mill Gulch is a shallow ravine several miles long and from 200 to 500 feet wide. The alluvium is composed largely of sand and soil with some subangular rock fragments. The depth to bedrock varies from 20 to 45 feet, averaging 30 feet. Bedrock is decomposed diorite. The ground has been prospected by 12 shafts sunk to bedrock at intervals of 500 feet along the gulch and a trench across the gulch. Some drift mining has been done along bedrock from the shafts. The average value of the placer material, according to Raleigh, is $1 per cubic yard. The values vary from a few cents per cubic yard near the surface to $6 per cubic yard on bedrock. About six inches of the bedrock itself will, in places, go as high as $30 per cubic yard. The gold is both fine and coarse. The largest nugget found in 1934 had a value of $12.63. Fineness of the gold averages 920. The gold in Mill and Triplett Gulches presumably has been derived from the quartz veins at the head of the gulches.

During the summer and fall of 1935 four men were leasing on the Mill Gulch placer and paying a royalty of 10 percent of the value of bullion recovered. The placer material was mined by drifting on bedrock. Value of gravel was $4 to $6 per cubic yard. The material as mined is slightly damp and it is necessary to dry it in the sun before concentrating it in the dry washer. Very little clay is present to interfere with dry washing. Power dry washers were used to recover the gold.

Sufficient water for placer operations can be obtained on the flat east of the placer ground. One well has been drilled to a depth of 310 feet in a search for artesian water, and the water level in this well is 70 feet from the surface. Raleigh sank a shaft on the flat nearer the placer ground than the aforementioned well, and a large flow of water was encountered at a depth of 40 feet. The lift from this well to the head of the Mill Gulch placer is about 500 feet.

IOWA CANYON DISTRICT

The Iowa Canyon District is about 16 miles north of Austin, Nevada. In recent years placer gold has been discovered in the canyon near the ranch of Joseph Phillips. A shaft was sunk in the canyon in an attempt to reach bedrock, but a large flow of water was encountered. Operations have not gone beyond the prospecting stage.

MUD SPRINGS DISTRICT

The Mud Springs District is eight miles northwest of Camp Raleigh. Placer gold was discovered in the District in 1907 by

Gus Fowler of Beowawe. The placers have been worked intermittently for a number of years on a small scale. It is reported that about $2,000 worth of placer gold has been produced. In the summer of 1935 four men, working with hand rockers, are said to have made wages.

Placer gold has been found in Mud Springs Gulch, Rosebud Gulch (tributary to the former), and Tub Springs Gulch, which is south of Mud Springs Gulch. Most of the placer mining has been done in Mud Springs Gulch, which is about four miles long and averages about 300 feet in width. About 30 shafts, 16 to 90 feet deep, have been sunk to bedrock along the course of the gulch. The average depth to bedrock is 30 feet. Some water is found on bedrock, and during the spring run-off the flow is quite large. The top six to eight feet of placer material is sand and soil, but below this stratum the placer gravel is composed largely of well-rounded small and large boulders. The placer material is cemented, so that dry washing is not feasible. The gold is coarse and angular at the head of the gulch, but becomes finer in proportion to the distance it has traveled downstream. The best values are concentrated near bedrock. Some of the gold is encrusted with quartz. In all probability the placer gold was derived from erosion of the quartz veins on Granite Mountain. When the gravel is worked with hand rockers a white concentrate, which is probably cerussite, is found with the gold. The gold has a fineness of 865 to 875.

In October, 1935, the A. O. Smith Corporation of Milwaukee sank several test shafts on a group of claims at the head of Mud Springs Gulch. After only a short time this work was discontinued.

LYON COUNTY
GOLD CANYON DISTRICT

The Gold Canyon District, also known as the Chinatown, Silver City, Devil Gate, or Dayton District, is in Gold Canyon on the east slope of the Virginia Range in western Lyon County. Placer gold was discovered by Abner Blackburn in 1849 in the sands of the Carson River at the mouth of Gold Canyon near Dayton. This was the first recorded discovery of placer gold in Nevada. From 1850 to 1857 a band of placer miners, whose number varied from 20 to 200, washed the Gold Canyon placers with rockers and long toms. The average wage made from this work is reported to have been about $5 per day per man. Water for working the placers was available during only a few months of the year; the rest of the time these early placer miners spent in prospecting and in making occasional forays for supplies into the Mormon settlements in Carson Valley. The District has been the scene of long-continued placer operations, and, although the total production of gold is unknown, it is undoubtedly quite large, judging from the extent of the old workings. During the last few years several small lots of placer gold were produced by drift mining and sluicing near the town of Dayton.

The Gold Canyon placers resulted from the disintegration of the lodes of Comstock and Silver City Districts. North of Dayton Gold Canyon spreads out into an alluvial fan, in which considerable gold has been found at varying depths up to 50 feet. Although the area is scarred with shafts and pits, from which ramify numerous gravel drifts, it is not improbable that gravels underlying the town of Dayton and the delta west of it may be rich enough to justify gold dredging. Southwest of Dayton the Carson River has etched a narrow but fertile valley, into the north side of which emerge ravines from the Comstock and Silver City Districts. Engineers have considered sampling this area for dredging, but the land is held closely by a number of small ranchers.

In 1920 a dredge was built in the District by the Gold Canyon Dredging Company, a subsidiary of the Metals Exploration Company, which at that time was working mines on the Comstock Lode. The placer holdings of the company lay between Silver City and Dayton and consisted of the Manuel King ground of 300 acres below Silver City and the Rae ground of 720 acres west of Dayton. The dredge began work on September 5, 1920, on a site two miles below Silver City, and it ran until April 5, 1923. About 3,000,000 yards of gravel were treated and, according to Mineral Resources of the United States, the gross production of bullion for the period was 14,625.3 fine ounces of gold and 7,482 fine ounces of silver, having a gross value of $309,750. According to J. H. Rae, Jr., of Dayton, 900,000 cubic yards were dredged on his holdings, from which about $105,000 in bullion was recovered, the rest was taken from the Manuel King ground. Considerable trouble was encountered in operating the dredge because of boulders and the loss of water from the dredge pond. At the upper end of the placer, old Chinese underground workings threatened to drain the dredge pond. In addition, the grade of the canyon is steep and it was necessary to construct a system of levees to hold the water. Large boulders encountered during dredging also proved a handicap, due to the fact that the dredge pitched violently in digging, causing loss of gold on the riffles.

The dredge was the close-connected bucket type designed to

handle 5,000 cubic yards per day. Each bucket held 9 cubic feet and the digging depth was 40 feet. The hull was steel, 46 feet wide and 108.5 feet long. The deck seams and the upper side seams of the hull were arc welded, while all the main construction joints were riveted. The total weight of the dredge was 900 tons and the cost of constructing it was $250,000.

The gold-saving equipment on the dredge consisted of a receiving hopper, a revolving screen 5 feet in diameter and 38 feet long, perforated by 1½-inch round holes, sluices, riffles, and a tailings stacker. The sluices and riffles had a total area of 3,000 square feet. The tailings stacker was 120 feet long and had a conveyer belt 34 inches wide. Water was supplied by a 10-inch centrifugal pump.

The dredge motors totaled 500 horsepower. A 23,000 - volt transmission line traversed the property and power was purchased from the Truckee River General Electric Company.

Water for dredging was purchased from the Virginia and Gold Hill Water Company which owns the water system that supplies Virginia City and Silver City. Water was brought in by flume and inverted syphon from Marlette Lake near the west shore of Lake Tahoe. Marlette Lake is 9½ miles in an air line southwest of Virginia City. The dredge required 35 miner's inches of water at $12 per miner's inch per month. The water was brought from the water line of the Virginia and Gold Hill Water Company to the dredge pond by 4,500 feet of 8-inch woodstave pipe and 3,800 feet of ditches.

In 1923 the dredge was dismantled, and several years ago it was sold to California interests that are now working it near Sacramento.

The Rae placer ground is on a terrace southwest of Gold Canyon, sloping toward the Carson River. A view of the placer ground is shown in figure 39. This ground has been tested by numerous shafts 28 to 90 feet in depth. None of the shafts have been sunk to bedrock as the gold values diminish beyond an average depth of 40 feet. According to J. H. Rae, Jr., past sampling by several companies indicate that the average value of the gravel is about 24 cents per cubic yard with present price of $35 per fine ounce for gold. Fineness of the gold, as calculated from the bullion recovered by dredging, averages 661.

From time to time the thought of placer mining the sands along the Carson River for the values lost in mill tailings is revived. These tailings were derived from the early-day mills used for the treatment of the Comstock ores. At one time it is reported there

were 150 mills in the District, many of them along the Carson
River. Virtually all of these mills were operated on a custom
basis, and many of them never produced anything except liti-
gation and assessments. These early - day mills employed the
Washoe process (amalgamation in pans heated by steam, using
quicksilver, salt, and copper sulphate for reagents), the Frieberg
process (chloridizing roasting with subsequent amalgamation in
barrels), Veatch process (same as Frieberg process, except that
steam tubs were used instead of the barrels), and the Patio process (amalgamation on an open floor with the aid of salt and copper sulphate). These early-day processes were crude as compared
with present metallurgical practice, and recovery was from 60 to
65 percent. The values lost in the tailings were deemed of minor

Figure 39. View of terrace placer across Gold Canyon, Dayton District.
White line is tailings pile from dredging operations along rim of Gold
Canyon.

importance. Although some of the tailings were impounded and
subsequently re-treated, a vast amount was diverted into the Carson River. It is reported that some $60,000,000 of values left
in the old tailings were sluiced down the canyons or deposited
directly into the Carson River.

In the early nineties a company, backed by Boston capitalists,
was organized to recover the values in these old tailings. Seventeen miles of "tailings" were located along the course of the
Carson River from Empire towards Dayton. Floured quicksilver could be panned or washed out along the banks, bed, and
flats of the river, and assays showed from a few cents to a dol-
lar or more per ton. Three dredges were constructed to work

these tailings, two of which were of the clamshell type, the third being a suction dredge, all powered by steam. Reported cost of the equipment was about $300,000. Attempts to work the tailings persisted over a period of eight years, when finally the rich conundrum was abandoned. This venture demonstrated that no dredging process could profitably overcome the expense of washing the vast amount of sand, mud, and gravel to recover the values lost in the tailings.

YERINGTON DISTRICT

The Yerington District, also known as the Ludwig or Mason District, is in the southern part of Lyon County in the vicinity of the town of Yerington. Although the District has been an important producer of copper since 1864, placer deposits were

Figure 40. Ancient river channel covered by recent placer gravel, Adams-Rice placer, Yerington District.

unknown until 1931, when J. S. Adams and Jeff Rice discovered gold in Big Canyon on the east slope of the Singatze Range.

The Adams-Rice placer comprises 400 acres of ground in Big Canyon on the west side of Mason Valley between Mason Pass and Gallagher Pass, about eight miles northwest of Yerington. The canyon is a little over two miles long and from 300 to 600 feet wide. Average depth to bedrock is 20 feet. Bedrock is chiefly andesite and porphyry. While testing the deposit with shafts, it was found that part of the placer gravel is underlain by an ancient river channel, which also carries gold. The gravel in the ancient river channel is cemented and composed largely of well-rounded boulders, while the later gravel is composed of small subangular rock fragments, pebbles, and sand uncemented and slightly stratified. Figure 40 shows the ancient river channel covered by later gravels where it has been exposed in an open cut on the property. Coarse and fine gold in equal amounts is found in the deposit. Some of the gold is sharp and angular and some is nuggety and worn by water action. It is believed that the angular gold was derived from the quartz veins at the head of the canyon while the water - worn gold was reconcentrated from the gravel in the ancient river bed that traverses the ridge above the placer workings. The fineness of the gold is from 895 to 945. The largest nugget found on the property had a value of $20. Black sands are present in amount averaging 20 pounds per ton of alluvium. Much of the gold consists of finely divided dust and flakes that has a tendency to float on water. The placer has been prospected and sampled by about 35 shafts sunk to bedrock. Each shaft is 4½ by 4½ feet in section. The cost of sinking these shafts by hand with windlass and bucket has averaged about $1 per linear foot of shaft.

Placer gold has been found also in the alluvial fan at the mouth of the canyon. Many claims have been staked out on this fan, but very little prospecting has been done.

The original discoverers, after some preliminary prospecting and leasing, sold the Big Canyon placer to a Reno group who made arrangements to work the placer ground with a power shovel. A general view of the placer is shown in figure 41. Pumping equipment was installed in the valley east of the placer and a washing plant was erected on the placer ground. Operations began in the fall of 1932.

In working the placer the overburden was removed with a power shovel and the gravel near bedrock was loaded by the shovel into Granby-type mine cars that held six cubic yards. The cars were hauled to the washing-plant bin by double-drum hoist geared to a 30-hp. motor.

From the bin a belt feeder carried the gravel to the trommel. The trommel was 20 feet long, 4 feet in diameter, and equipped with ¼- and ½-inch mesh screen. About 60 percent of the gravel passed through the ¼-inch mesh screen. The trommel oversize was carried to the waste pile by a conveyer belt 150 feet long and 22 inches wide, inclined about 12 degrees. The undersize was fed to sluice boxes 16 x 16 inches in section, built in two lengths, each 75 feet long. A 5-inch centrifugal pump connected to a 25-hp. motor was used to elevate the gravel from the first to second length of sluice. The gravel was elevated because it was necessary to dispose of the tailings in a side gulch. In the first length of sluice Hungarian riffles were used, and cocoa matting and woven-wire screen were used in the second length.

The capacity of the plant was 60 cubic yards per hour. It operated only a short time, when the property was taken over by the Apex Mining Company. This company worked for a short time in 1934 and 1935, removed about 12 feet of over-burden with a Sauerman dragline, and used a power shovel to mine about eight feet of gravel on bedrock. A portable placer machine, mounted on caterpillar treads, was used to recover the

Figure 41. View of Adams-Rice placer, Yerington District.

gold. The placer machine was moved with the power shovel. In 1934 a test run of the machine is reported to have recovered about 100 ounces of gold in a little over three months. After working a short time in 1935, the company discontinued operations. It is said that the machine could not be maneuvered so as to follow the meanderings of the pay streak, and, in addition, the problem of disposing of the tailings proved a drawback. Power was purchased from the Sierra Pacific Power Company.

Water is obtained from a 165-foot well in the flat east of the placer. The water level in the well is 20 feet from the surface. From the well the water is pumped to a tank, 12 feet in diameter and 7 feet high, by a 7-inch Pomona deep-well pump, belt

driven by a 15-hp. motor.   From this tank the water is fed to
two Gould triplex pumps, 3½ by 8 inches and 4½ by 8 inches,
belt driven by motors of 30 and 38 horsepower.   The pipe line
is 4 inches in diameter, 2 miles long, and has welded joints.
The pumping lift to a 105,000-gallon wood-stave tank above the
placer ground is 600 feet.   The capacity of the pumping plant
is 155 gallons per minute.   When the first placer plant was
erected on the property, provision was made for reclaiming some
of the water in a settling pond.

About two miles southwest of the Adams - Rice placer on
Lincoln Flat in the Singatze Range, R. J. Penrose and two
partners have, for several years, been prospecting a Tertiary
river channel for placer gold.   This Tertiary channel has been
traced on the surface for a distance of six miles by Penrose and

Figure 42.   Prospecting layout at Penrose placer, Yerington District.

partners, and undoubtedly it belongs to the same "dead river"
system as the channel found on the Adams-Rice placer.

A shaft (see figure 42) has been sunk in the channel to a
depth of 120 feet, and several hundred feet of lateral work has
been done from this shaft without encountering any rich gravel.
The river channel has been disturbed considerably by faulting.
The average value of the gravel taken out during prospecting
operations is reported to be 20 cents per cubic yard in gold.

The gravel consists of numerous well - rounded boulders and
finer material tightly cemented, so that blasting is necessary.
The material is heavily stained with iron oxide.   The gold found

in prospecting is mostly fine and is associated with a considerable amount of black sand.

No water is available in the vicinity of the prospecting operations and samples, after drying, are run over a power-driven dry concentrating table, shown in figure 43.

MINERAL COUNTY
HAWTHORNE DISTRICT

The Hawthorne District is in the vicinity of the town of Hawthorne in central Mineral County. In 1909, during the mining excitement that followed the discovery of rich veins in the Lucky Boy District, placer gold was discovered near Mount Grant. The placers received little attention until 1932, when it was discovered that profitable placer values occurred five or six feet below the surface in a bed of volcanic ash. From 1932 to

Figure 43. Power-driven dry concentrating table, used in sampling at Penrose placer, Yerington District.

the present, C. B. Murray of Reno has been working a placer deposit at Laphan Meadows near Mount Grant.

The placer material is excavated with a dragline. About four feet of overburden is removed first and then about two feet of gravel above bedrock is transported to a portable sluice by the dragline. Water for sluicing is obtained from several springs in the vicinity. The supply is limited and the water used in sluicing is reclaimed. Maximum capacity of the plant is 100 cubic yards per day. Five men are employed.

The gravel is uncemented detrital material with few boulders.

The gold is mostly coarse. The largest nugget found in 1935 had a value of $30.

In 1935 several boys from the Hawthorne CCC camp discovered placer gold in Baldwin Canyon several miles south of the Murray placer. The new discovery has not been prospected sufficiently to determine its importance.

PINE GROVE DISTRICT

The Pine Grove District, also known as the Rockland or Wilson District, is on the east flank of the Smith Valley Range in western Mineral County, 20 miles south of Yerington. The District was discovered by William Wilson in 1866 and about $9,000,000 has been produced from the lode deposits. Placer gold has also been found in the District, but the production has been small.

Some placer mining has been done on the slopes of Sugar Loaf Mountain near the mouth of Pine Grove Canyon. Several tunnels and shafts were driven by Victor Barnard of Yerington in 1926, but the work in general proved disappointing. The placer material is detritus from the veins in the Wheeler and Wilson mines. Large boulders occur in the placer material.

RAWHIDE DISTRICT

The Rawhide or Regent District is in the northwestern part of Mineral County near the Churchill County border. The lode deposits were discovered in 1906 and the placer deposits a short time after. The total production of the placers has been about $250,000, all of which was obtained by small dry-washing operations, and most of it in 1913 when the placers were worked intensively. It is said that as many as 100 men were engaged in placer mining in the District in the early days, some of them earning as much as $30 per day with hand dry washers. In recent years there has been a small amount of dry washing in the District by itinerant placer miners, but the returns from this work have been less than wages.

The best placer ground is an area about half a mile wide and one mile long on the southeast slope of Hooligan Hill (see figure 44). Numerous shafts, averaging about 15 feet in depth, in this area attest the activity in former days. No placers have been found on the opposite slope of Hooligan Hill. Gravel is angular rock fragments, sand, and soil, with few large boulders. Bedrock is presumably rhyolite. The largest nugget found in this area had a value of $70.

The placer ground on Hooligan Hill slopes toward a canyon several miles long. At the south end of this canyon the gravels spread out to form an alluvial fan, in which placer gold has been found. The gravels on the fan are 90 feet deep in places and the best values are concentrated on the bedrock. About 30 shafts have been sunk on the fan, several of which have not reached bedrock. Placer mining was done at several of the shafts in 1931 and 1932, a power-driven placer machine being used to recover the gold from the gravel extracted by drift mining. The first five feet of gravel above the bedrock, worked by dry washer, is said to have contained as much as $5 per cubic yard in places.

In 1930 the Idaho Dredging Company of Boise, Idaho, obtained a bond and lease on 1,800 acres of ground in the District and

Figure 44. Old placer workings on southeast slope of Hooligan Hill, Rawhide District.

began sampling with the object of dredging the ground if sampling showed enough gold. Water is available in the valley east of the town of Rawhide. After a short time the property and the sampling operation were taken over by the Hammond Engineering Company. The latter company continued sampling until May, 1931, when operations ceased and the property reverted to the original owner. The reason for the relinquishing of the option by the two companies is not stated, but in all probability it was the erratic distribution of the gold which resulted in low average values. It is said that sampling operations showed that the bedrock is very irregular and in places comes close to the surface.

A number of placer claims were relocated in 1935 southeast of Rawhide, but little work was done on them.

TELEPHONE CANYON DISTRICT

The Telephone Canyon District is on the west side of the Pilot Mountain Range in southwestern Mineral County, five miles east of Sodaville. Placer gold was discovered near the mouth of the canyon in 1931, and a number of placer claims were located. Placer operations have not proceeded beyond the prospecting stage, and not more than a few ounces of placer gold have been produced. Telephone Canyon is a narrow, steep canyon about six miles long. The difference in elevation at the mouth and at the head of the canyon is nearly 3,000 feet.

In 1931 a shaft was sunk to a depth of 70 feet to bedrock near the mouth of the canyon by Walter Fancher of Manhattan and George Mannington of Tonopah. Water was encountered at 40 feet and a pump was installed, but the flow of water encountered near bedrock was too great for the capacity of the pump and the work had to be abandoned.

The gravels exposed in the banks near the mouth of the canyon are made up largely of fine, unassorted pebbles and sand, with no large boulders. Gold is all fine and associated with much black sand. In 1935 twelve ½-cubic-yard samples were taken from the banks, and fire assays of the concentrates obtained from these samples gave an average of $2.12 per cubic yard. Apparently the gold in the placers was derived from the network of small veins and stringers in quartzite and porphyry in and near the Belleville mine at the head of the canyon.

NYE COUNTY
BEATTY DISTRICT

The reported discovery of placer gold in this District in 1924 caused some excitement, and a number of claims were located along the Amargosa River (dry for most of the year) above and below the town of Beatty in southwestern Nye County. A little placer gold was found, but the reported discovery proved to have been of little or no economic importance.

CARRARA DISTRICT

The Carrara District is in southern Nye County eight miles south of Beatty, Nevada. Placer claims have been located in the District in the past, but no placer gold has ever been produced. No doubt some gold has been eroded from the veins in limestone on the west slope of Bare Mountain, but the gravel wash at the base of the mountain is deep and the gold is too widely disseminated through the gravel to be of any importance.

CLOVERDALE DISTRICT

The Cloverdale District, also known as the Golden or Republic District, is in the northwestern part of Nye County about 40 miles northwest of Tonopah, Nevada. Placer gold was discovered in the District in 1906, about four miles east of the Cloverdale Ranch. In the years following the discovery, small placer operations were conducted, but the amount of placer gold mined has been small. On account of the scarcity of water in the vicinity, placer mining has been restricted to dry washing or other methods that require little water.

In 1931 a Los Angeles group prospected placer ground in the District by 26 shafts 20 to 50 feet in depth. Some of the shafts encountered water. This project, covering 1,760 acres and including the whole of Cloverdale Canyon, was to be developed with a dragline equipment to handle 5,000 cubic yards per day. Machinery of special design for recovering the gold was delivered to the ground in 1931, but for some reason the project never reached the producing stage.

JOHNNIE DISTRICT

The Johnnie District is in southeastern Nye County on the east slope of the Montgomery Mountains. Lode deposits were discovered in the District in 1903. Placer gold was discovered in the District west of the Johnnie mine in 1921 by Walter Dryer, and a short-lived boom ensued. It is said that this discovery of placer gold was antedated by the Mormon pioneers, who discovered placer gold in the vicinity in the early days. The amount of placer gold taken from the District is not known, but in all probability it does not exceed $20,000.

The placer area follows the trend of the main lodes, extending from a point two miles north of the Johnnie mine to four miles south of the old Congress mine, a distance of about ten miles, with an average width of about one mile. Some placer gold has also been found on the west slope of the Montgomery Mountains.

During 1935 about 20 men worked the placers in the District. Most of these were itinerant placer miners who worked only a few weeks or at most several months. In November, 1935, six dry washers were working in the gulches below the Congress mine. Water is scarce in the vicinity, and all placer gold is won by working the gravels in hand-operated dry washers.

The Johnnie placers have been derived from the disintegration of the quartz veins in the vicinity. The principal placer diggings are in the bottom of the gulches and the tributary side draws. The depth of the gravel varies from 3 feet to 25 feet.

The best values are concentrated on bedrock. The gravel consists largely of small angular rock fragments mixed with soil and some clay. Some boulders are present in the detrital material. About six inches of material on bedrock will run from $6 to $30 per cubic yard. The thickest pay streak was eight feet in the main Johnnie gulch at 25 feet below the surface. The pay channels split and reunite, and their width varies considerably. Schist or shale bedrock, having upturned edges that form natural riffles, is the most satisfactory condition for concentration.

Some of the country rock below the Congress mine is covered with beautifully dendritic markings. These markings are usually mistaken by the placer miners for impressions of plants, but they

Figure 45. Hand-operated dry washer used in Johnnie District.

are of inorganic origin, having been formed by percolating waters carrying oxide of manganese, which has assumed the dendritic or treelike form.

The gold is rough and angular, frequently encrusted with quartz. The fineness of the gold averages about 880. The largest nugget found in the District in 1935 weighed 3½ pennyweights.

In 1935 most of the placer miners were leasing on placer ground, owned by Matt Kusick, in the gulches below the Congress mine, and paying a royalty of 10 percent of the gross returns. Figure 45 shows a dry washer, operated by B. R. Beach and his wife, typical of the others in the District. The gravel is hoisted by hand windlass and 2-cubic-foot bucket. In drift mining the bedrock is cleaned thoroughly with wire brush and whisk broom. An average of one cubic yard of gravel is mined and treated per day, and an average return is about one ounce of gold per week. Dry washing is effective in recovering the gold as the material contains little clay and is dry. A considerable amount of black sand and pyrite is found in the dry-washer concentrate. This concentrate is cleaned by hand panning, as shown in figure 46.

In former years hand rockers were used to some extent, but for the greater part of the year the water available in the District is just about sufficient for domestic purposes only. This water is piped to the camp of the Johnnie placer from Horsehutem Springs in the Charleston Range across Pahrump Valley

Figure 46. Cleaning dry-washer concentrate by hand panning, Johnnie District.

to the east. The gravity pipe line is a little over four miles long.

MANHATTAN DISTRICT

The Manhattan District is in the southern part of the Toquima Range in Nye County, about 50 miles north of Tonopah by automobile road. The altitude of the District is about 7,000 feet. The District has an arid climate, which is mitigated by the high mountains to the north and east. The precipitation is irregular, and in some years the rainfall for a whole year may be concentrated in a single storm.

The date of the discovery of placer gold in the District is vague. In 1905 the Dexter Mining Company sank a well for water in Manhattan Gulch; colors were found in the well, but no attention was given to this discovery. In the summer of 1906 nuggets were found at the surface above the gulch by a miner named Burns, and the attention of the miners was diverted from lode mining to the possibilities of placer mining. By 1908 drift mining of the deep gravels and dry washing of surface material was well under way, and from that time to the present there has been placer mining in the District.

From 1907 to 1930 placer gold to the value of $1,325,173 has been mined. In 1907 the amount was $1,256, and it has gradually increased until it reached a maximum of $170,499 in 1912, after which it slowly declined due to the exhaustion of the richer portions of the deposits which it was possible to work economically by hand methods. The bulk of the gold was won by drift mining the gulch gravels. It is reported that in the early placer work the gravel removed above bedrock averaged from $5 to $30 per cubic yard.

The well - rounded topography in the vicinity of Manhattan indicates extensive erosion. The geology of the Manhattan placers has been described by Ferguson.[18] As interpreted by Ferguson, the placer gold deposits in the District consist of three types—the older bench gravels, which are remnants of stream erosion; the buried gravel of gulch; and the recent wash deposited on hillsides and in dry stream courses.

Manhattan Gulch has been drift mined for a length of about five miles. The average width of the gulch is about 300 feet. The grade of the gulch, which is approximately the same as that of the bedrock, is about 4 percent. The slope of the rim rock varies from 30 to 50 percent. The gulch bedrock is composed mainly of schist and shale. The depth of the gravels varies from 20 to 100 feet. The gravel consists of subangular material made up of limestone, quartz, shale, schist, and quartzite, with no large boulders. About 60 percent of the gravel is larger than one inch and the rest is sand, gravel, and small flat pieces of shale and limestone. The gravel is compact, so that in sinking shafts the only timber necessary is a collar set. In most of the shafts the bedrock pay streak is from two to four feet thick. As with most of the Nevada placers, water in the immediate vicinity is scarce and insufficient in amount to permit working the deep gravel on an extensive scale, such as hydraulicking or dredging. There is a flow of water along bedrock in Manhattan Gulch that appears to follow the south side of the gulch. This flow is estimated to be about 120 gallons per minute, and it has been utilized in treating the gravels by sluicing at the surface.

[18] Ferguson, Henry G., Geology and Ore Deposits of the Manhattan District, Nevada: U. S. Geol. Survey Bull. 723, 1924, pp. 117–133.

Fine and coarse gold is found throughout the gulch. For the most part the gold shows very little the effect of abrasion and is usually aborescent in shape. Nuggets up to an ounce in weight are not unusual. In 1934 a nugget was found by one of the lessees in Manhattan Gulch which sold for $125. This value was a little more than its intrinsic worth. The fineness of the gold varies from 705 to 735, averaging about 732. The fineness of the gold increases with the distance downstream. The concentrates obtained in the clean‑ups contain black sand and barite. A small amount of cinnabar is found also. One operator reports that he found a small amount of cassiterite at the lower end of Manhattan Gulch. In places the black sands assay as high as $200 per ton.

Most of the placer mining has centered in Manhattan Gulch, although some has been done in Giffen Gulch south of Manhattan Gulch and in Slaughterhouse Gulch to the north. The bench and hillside gravels have not been worked as intensively as the gulch placers.

In the early days of placer mining, shafts were sunk to bedrock at intervals of about 300 feet, as it was deemed uneconomical to exceed that distance. After the pay streak had been developed, the gravel was taken out by drift mining, retreating toward the shaft after the manner of long‑wall coal mining. The bedrock contains small crevices that acted as natural riffles, and this bedrock was removed to a depth of about a foot. Tramming was done with wheelbarrows or small mine cars running on track. Boulders too large to be hoisted readily were left behind as the work progressed.

In washing the gravels, various types of equipment were used. Among the best of the early washing plants was one installed by Thomas Wilson. The main feature of the plant was the hoisting equipment, which consisted of an endless-chain elevator about 20 inches wide. This raised the gravel from an underground feed bin of 35-cubic-yards capacity constructed in bedrock. The depth of the shaft was 65 feet. The chain elevator discharged into a 25-cubic-yard capacity storage bin that was built 25 feet above the surface. From the storage bin the gravel was sluiced whenever sufficient yardage had accumulated. From the chute of the storage bin the gravel discharged directly into a revolving screen trommel with 1½-inch punched holes. The oversize was discharged from the lower end of the trommel into a chute, which diverted it to a waste pile. A steady stream of water played on the trommel to assist in disintegrating any

cemented gravel and to wash the loose clay from the gravel.
Directly below the trommel was a shaker sluice, which consisted of a box similar to a sluice box, but with deep riffles
crosswise every two inches of its 12-foot length.   By a system
of pivots, the box was rapidly shaken vertically and laterally
with a play of about three inches lengthwise and one inch up
and down.   At the lower end of the shaker three small copper
amalgamation plates were fixed.   The gravel from the shaker
sluice passed over the plates, and any fine, light gold that did
not settle between the riffles was amalgamated with the quicksilver on the plates.   Stray amalgam from the plates was caught
in the riffles below.

Below the plates a line of sluice boxes, all containing riffles,
some of which were placed transversely and some longitudinally, extended for some distance.   The small amount of gold
that escaped the shaker and amalgamation plates was caught
in its travel down the sluice boxes.   The shaker sluice box
caught about 90 percent of the gold recovered.   The power for
the elevator, trommel, and shaker sluice was furnished by a
15-hp. motor.

In 1935 about 30 men were engaged in placer mining in the
District, using sluice boxes, machines of special design, and dry
washers to recover the gold.   Most of the dry washing was
confined to reworking the old dumps in Manhattan Gulch.   As
the ground is held in large tracts by a few individuals, placer
operations are conducted largely under the leasing system.   The
customary royalty paid in the District was either 10 or 15 percent of the gross returns.   Placer miners in the District averaged wages for their efforts.

The surface layout on the placer lease operated by W. R.
Amidon, Fred Walker, and Richard Walker in Manhattan Gulch
is shown in figure 47.   The leased ground consists of a block
300 feet square.   The royalty paid is 15 percent of the gross
returns.

The shaft is 42 feet deep, and is inclined 75 degrees.   Hoisting
is done with a small bucket holding $\frac{1}{10}$ cubic yard, sliding on
skids, and a home-made hoist driven by a 5-hp. motor.   In order
to eliminate the services of a hoist man at the surface, the hoist
is mounted at the bottom of the shaft.   At the surface the bucket
dumps into a bin having a capacity of six cubic yards, by means
of the ordinary self-dumping arrangement, whereby the lugs of
the bucket drop into slots cut into the skids.   Tramming is done

with a bucket placed on a small flat car that runs on 12-pound rails.

In mining, four feet of the gravel above bedrock is removed, together with from 6 inches to 1½ feet of the slate bedrock. The best values are found directly on and in the crevices of bedrock. Average value of the gravel mined is $3 per cubic yard. No water is found on bedrock. The few boulders encountered in mining are left behind in the old workings.

The washing plant consists of a shaking screen and sluice box. The flow of gravel from the bin to the shaking screen is controlled by rack and pinion gate. The shaking sluice is about 6 feet long, 14 inches wide, and about 10 inches deep, and is

Figure 47. Surface layout on Amidon-Walker placer lease, Manhattan Gulch, Manhattan District.

made of ³⁄₁₆-inch sheet iron with 1¼-inch-diameter holes. The screen is shaken longitudinally at the rate of 30 strokes per minute by an eccentric arm actuated by belt drive from a 3-hp. motor. Length of stroke is 4 inches. Below the shaker is a wooden trough leading to the sluice. Several riffles are in the trough to catch the coarse gold. The oversize from the screen is diverted to a 1-ton mine car for tramming a short distance to the waste pile.

The sluice box is 45 feet long and 1 x 1 foot in section with a slope of 1½ inches per foot. Transverse riffles, 2 x 2 inches ,with 2-inch space between successive riffles are used in the sluice. The capacity of the washing plant is about 9 cubic yards in 1½ to 2 hours. Fifty percent of the gold recovered

5

is found in the trough below the shaker, and the other 50 percent in the first 15 feet of sluice.   No mercury is used in the
sluice boxes.   Clean-ups are made every shift.   The concentrates
obtained in sluice-box clean-ups contain considerable black sand
and fine barite.   The concentrates are cleaned by hand panning.
Fineness of the gold is about 715.   Up to the time of the writer's
visit the lessees had been working two months and the average
clean-up was a little better than an ounce of gold per day for this
period.

The capacity of the plant is a little more than four cubic yards
per man-shift of seven hours.

Water for sluicing is obtained from the tailings of the Matt

Figure 48. Placer plant at Baisden-Waters lease, Manhattan Gulch,
Manhattan District.

Kane mill located above the placer plant.   The tailings water is
impounded in a shallow reservoir dug in the gulch, and after
settling the water is pumped to the shaker sluice by a 2½-inch
centrifugal pump connected to a 7½-hp. motor.   Pumping capacity is 150 gallons of water per minute.

Electric power is purchased from the Nevada - California
Power Company at a cost of $15 per month.

The placer plant on the S. E. Baisden-Morris Waters lease in
Manhattan Gulch is shown in figure 48.   The ground held under
lease is 300 feet square.   The lease is good for 6-month periods,
and the royalty is 15 percent of the gross returns.

Entry to the mine is made by vertical shaft sunk 30 feet to
bedrock.   Hoisting is done with a ⅕-cubic-yard bucket and single

drum-hoist belt driven by a 5-hp. motor. The hoist is at the surface. Tramming is done with a bucket placed on a mine truck that runs on 12-pound rails.

In mining, four feet of the gravel above bedrock is removed, as well as about one foot of the shale bedrock. The pay channel is 25 feet wide. About 50 percent of the gravel mined is minus 1-inch mesh.

The regular procedure in operating is to mine and hoist gravel for five hours, and wash gravel and make a clean-up two hours in the afternoon. The average yardage handled per day with two men is seven cubic yards. The operators state that the gravel must run at least $2 per cubic yard in order to make wages.

The bin above the shaft has a capacity of seven cubic yards. The shaking screen and sluice box used are essentially the same as those at the Amidon-Walker lease.

Three 5-hp. motors are required to operate the placer equipment—one for the hoist, another for shaking screen, and a third for the 2-inch centrifugal pump. Water for sluicing is obtained by impounding the tailings from the Matt Kane amalgamation mill. The water from the placer plant is reused. Water consumption is about 100 gallons per minute.

The cost of the placer plant is about as follows:

Transformer and fuse blocks (used)	$300
Three 5-hp. motors (used)	150
Hoist (used)	50
Headframe, sluice, shaking screen, bin	250
Total	$750

Labor for installation is included. Power costs about $15 per month.

In 1935 the operators recovered $2,800 in seven months. Five of the seven months were spent in working on ground other than the present lease. Fineness of the gold averaged about 710. The largest nugget found in 1935 weighed 1½ ounces.

The lease on the Patrick placer ground at the lower end of Manhattan Gulch is operated by Jack Farrell, Herman Smith, and Art Smith, who pay a royalty of 10 percent of the gross returns.

Entry to the underground workings is made by vertical shaft sunk 35 feet to bedrock. Hoisting is done with hand windlass and a bucket of 2-cubic-foot capacity. At the time of visit, the lessees were engaged in prospecting by drifting along bedrock.

The average grade of the gravel taken out during prospecting was $1.25 per cubic yard. The gravel must average better than $2 per cubic yard in order to net wages.

Tramming is done with a Klondyke wheelbarrow and bucket. This type of wheelbarrow is made of 1½- or 2-inch iron pipe and pipe fittings. The pipe framework is constructed so as to support the bucket. By using the same bucket for tramming and hoisting, transfer of material is avoided. The slate bedrock is cleaned with a wire broom, as the best values are found on bedrock. An average of five feet of gravel above bedrock is removed in prospecting operations. With one man at the surface and two men underground, the amount of gravel handled averages about seven cubic yards per shift.

The gravel is washed in a G. B. portable placer machine. The

Figure 49. G. B. portable placer machine and surface layout at Gaylord lease, Manhattan Gulch, Manhattan District.

placer machine and surface equipment are similar to those at the plant on the Gaylord lease shown in figure 49. The placer machine is belt-driven by a 3-hp. motor. The plus 1-inch product is screened out and discarded as waste before washing.

Water for placering is obtained from the Matt Kane mill tailings. From a small reservoir constructed in the gulch, the water is pumped to a tank 6 feet in diameter and 8 feet high located above the placer machine. The pump is a 1½ - inch centrifugal driven by a 5 - hp. motor. Water consumption is approximately 80 gallons per cubic yard of gravel. Electric power is purchased at a cost of about 4 cents per kilowatt-hour. Power consumption averages nearly $20 per month.

During part of 1934 and part of 1935 the Natomas Company of California sampled placer ground in and at the mouth of Manhattan Gulch. Three Keystone drills were used in sampling. The ground drilled included the Donald placers, comprising 3,400 acres on the alluvial fan and two miles of gulch gravel at the lower end of Manhattan Gulch. This ground has been sampled by nearly 200 drill holes and shafts, and the average value is reported to be sufficiently high to justify dredging operations. The deepest hole drilled was 170 feet to slate bedrock on the alluvial fan.

Working the gravels by dredging depends largely on the water supply. A surface and subsurface flow of water is available at Peavine Ranch, eight miles to the east. Peavine Creek drains a mountainous area of some 110 square miles, and the flow of water, according to William Donald, is about 1,500 gallons per minute. San Pablo Creek, several miles north of Peavine Creek, has a smaller flow of water. This creek drains an area of about 50 square miles.

ROUND MOUNTAIN DISTRICT

The Round Mountain District is in central Nye County, on the east side of the Toquima Range, about 15 miles north of Manhattan. Tonopah, 60 miles south, is the nearest railroad point and supply center. The District is served by daily automobile stage from Tonopah. The town of Round Mountain has an elevation of about 6,300 feet.

The principal lode deposits were discovered by Louis D. Gordon in 1906, and in the same year Thomas Wilson discovered placer gravel. The total placer yield up to and including the year 1932 is $1,295,920, which was obtained from 1,383,400 cubic yards of gravel, an average of 94 cents per cubic yard. Of this amount, the lessees produced, from 1906 to 1915, $442,940 from 272,240 cubic yards of gravel, an average of $1.63 per cubic yard. Hydraulic mining produced the remainder. From 1933 to 1935, inclusive, 98,420 cubic yards of gravel were treated by hydraulicking, and in 1934 and 1935, 82,282 cubic yards were treated in a mechanical placer plant. The average value of the latter was a little less than $1 per cubic yard.

The mountain from which the camp derives its name rises several hundred feet above the surrounding terrain. The principal lode and placer deposits are on Round Mountain. The more productive veins are the Los Gazabo, Keane, and Placer. From the erosion of these veins and the stringer zone lying between them the main placer deposits have been derived.

The main placer deposit is on the south and west slopes of Round Mountain. It is of the residual type, none of the gold having traveled more than several hundred feet from its source. The depth of the gravel varies from a few feet up to 50 feet. The gravel consists of about 20 percent coarse angular boulders of rhyolite and 80 percent sand and gravel, slightly stratified. It is cemented by a limy deposit into a hard conglomerate particularly near bedrock. Some gypsum is also found in the deposit. The rhyolite porphyry bedrock is very uneven, so that considerable gold is retained in the crevices, which necessitates considerable hand work to recover it.

Although gold is disseminated throughout the gravel from the surface down, the best values are found on bedrock. The increase in the value per cubic yard, obtained by prospecting the ground sloping toward Big Smoky Valley, is shown below:

DEPTH, FEET	VALUE PER CUBIC YARD
0–10	$0.202
10–20	.233
20–30	.642
30–40	1.70
40–50	2.89
50–60	3.55

The gold is angular and coarse and shows no evidence of having been transported. Nearly all the nuggets are encrusted with quartz or siliceous limonite. The gold averages about 635 fine, being alloyed with silver. The proportion of gold to quicksilver in the amalgam recovered in sluicing operations is about 60 percent. The concentrates, in addition to the gold, contain some finely divided magnetite and a little scheelite.

Placer mining was first carried on in the District in 1906 with dry-washing machines. A description of the early operations with dry washers has been given by Packard.[19] The machines used at that time were essentially the same as those in general use throughout the Southwest.

In operating these dry washers the large rocks were picked out by hand and thrown to one side. The gravel was then thrown against a screen having one-inch openings. The undersize was shoveled upon the dry washer screen, which had $\frac{1}{4}$-inch openings. The oversize was discharged to waste over a piece of sheet iron that projected two feet beyond the lower end of the machine, and the undersize was diverted to the head end of the machine,

---

[19] Packard, George A., Round Mountain Camp, Nevada: Eng. and Min. Jour., vol. 83, 1907, p. 151.

where it was fed upon a frame covered with coarse, heavy cotton cloth, across which wood riffles were placed about four inches apart. This frame formed the upper side of a bellows operated by turning a crank. The puffs of air through the cloth agitated the gravel, and, aided by the slope of the frame, discharged it at the lower end.

Two men were required to operate a single machine; one man turned the crank that actuated the bellows, and the other shoveled the gravel. When about 100 shovelfuls had passed over the machine, the frame was removed and replaced by another. The concentrate, consisting of black sand, gold and some gravel, retained on the riffles was brushed into a tub. Later the accumulated concentrates were put over the machine a second time, and the tailings which carried from $40 to $60 in gold per ton were sacked for shipment. The concentrate from the second washing was cleaned in a gold pan and the black sand removed by a magnet. One machine, with two men working about 10½ hours per day, handled about 17 tons of gravel. The depth of the gravel worked was from one to six feet. The dry washer recovered about 70 percent of the gold. While no definite figures are available as to the gold content of the gravel, it is believed that the material handled by the first dry washers contained from $2 to $10 per cubic yard. Inasmuch as the dry-washing operations were confined to the gravel in gulches, small basins, and other natural concentration areas, the values per cubic yard given are not unreasonable. Later, the ground directly below the site of dry-washing operations was hydraulicked by lessees, who treated 100,000 cubic yards that gave an average return of $2 per cubic yard.

From 1906 to 1915 all placer mining was done by lessees. In 1907 the Round Mountain Hydraulic Company was organized and water was piped from Jefferson and Shoshone Creeks in a pipe line 12 to 15 inches in diameter and several miles long.

In 1914 the Round Mountain Mining Company, the predecessor of the Nevada Porphyry Gold Mines, Inc., began the construction of a pipe line from Jett Canyon to the placer deposit. Hydraulic operations began on July 13, 1915, and continued to September 3, when the flow of water in Jett Creek had declined to such an extent that operations for the year had to be discontinued. In this first season, 18,150 cubic yards of gravel were mined which yielded $36,414 in bullion. The operating cost was $11,058, leaving a net profit of $25,356. Two dollars per cubic yard were recovered, and the total cost per yard was 61 cents.

As the water supply from Jett Canyon was inadequate for hydraulicking on such a large scale, the water rights of Jefferson, Slaughterhouse, and Shoshone Canyons were purchased by the Round Mountain Mining Company from the Round Mountain Power and Water Company. This enabled the company to handle a much greater yardage by lengthening the placer season and reducing operating costs. Placer operations have been carried on every year since, the length of the operating season depending on the amount of water available.

The pipe line from Jett Canyon is 45,336 feet long. The first section, beginning at the intake in Jett Canyon, consists of 14,000 feet of riveted steel asphaltum-dipped pipe, 30 to 15 inches in diameter. The second section consists of 28,000 feet of lap-welded steel pipe from $\frac{1}{4}$ to $\frac{5}{16}$ inch thick. This lap-welded pipe was necessary because the pipe line in the floor of the valley forms an inverted syphon 1,142 feet lower than the intake. The maximum static head on the pipe line in the valley floor is equivalent to a pressure of 495 pounds per square inch. The third section of the pipe line is 3,336 feet long, and is made of riveted steel asphaltum-dipped pipe 15 inches in diameter. The entire pipe line is buried 42 inches deep for protection against freezing.

Of the riveted pipe, 7,550 feet is equipped with slip joints where the pressure is light; the remainder is plain end with bolted steel, forged couplings. The pipe line traverses Jett Canyon for 10,000 feet and then continues in a straight line across the valley to the mine. It was designed with a safety factor of five, and is equipped with drains, air valves and pressure-relief valves. Exclusive of the Jett Canyon water right, which cost $28,410, the cost of the pipe line was $150,557. The difference in elevation at the intake of the pipe line and at the placer banks is about 650 feet.

After the high water season had passed in 1916 and 1917, it was the practice to close the valve at the head of the Jett Creek line and to allow the water to back up in the line for several hours, after which the valve was opened and sufficient water obtained for short periods of hydraulicking. Despite the safety factor on the line, this did not seem very good practice, so, in the latter part of 1917, an earth dam was built above the intake in Jett Canyon. This dam cost $3,000 and permitted the impounding of sufficient water to provide a full head for a large monitor for eight hours. The expense of the extra shifts was thereby eliminated.

In order to secure more continuous operation at Round Mountain it was necessary to provide additional impounding facilities. Until 1920 the pipe lines from Jett and Jefferson Canyons were connected directly to the giants. The additional impounding facilities were obtained in 1921 by building a small reinforced concrete dam in a gulch about one mile east of the placer banks. This dam is 22 feet high, in the center of the gulch, and impounds sufficient water from the two pipe lines during the low-water periods to permit the continuous operation of one large monitor for seven hours. The difference in elevation between this dam and the placer banks is 350 feet.

The first sluice, built in 1915, was 2 feet wide, 2 feet deep, and set at different grades. This sluice proved to be too small and too far above bedrock. The site of the new sluice was determined by drilling an appropriate number of holes to bedrock along the course of the proposed sluice. A new sluice, built in 1921, was 3 feet deep, 3 feet wide, and 5,000 feet long, with a uniform grade of 4 inches per 12 feet. The larger capacity and greater efficiency of the new sluice reduced the cost per cubic yard of moving the gravel. It is constructed of 12-inch plank with 2 x 4-inch lining boards. The supporting sills and side members are 4 x 4-inch timbers with 1 x 6-inch braces. The sides are rough lumber, but the bottom is built of clear plank 18 inches wide, surfaced on four sides in order to reduce the loss of quicksilver and finer particles of gold. This sluice had outlived its usefulness, and after the placer season of 1932 was over a new one was constructed to serve another part of the placer ground that had been drilled and sampled. The new sluice is the same size.

In early hydraulicking it was the practice to use a giant at the lower end of the sluice for piping away the tailings as they accumulated. As water is the factor that controls the amount of material that may be handled, a scheme was devised whereby the use of the tailings giant was obviated. A system of branch races was installed at the lower end of the sluice with gates that permitted diverting the flow of water and gravel from one branch to another. One branch was extended while the tailings were discharged through another. This arrangement permitted all the water to be used for hydraulicking.

The first riffles used at the Round Mountain placer were made of pine blocks about six inches in length set in the bottom of

the sluice in rows. Each row was separated by a riffle stick
of 1 x 4-inch board. With this arrangement of the riffles a
section from 50 to 100 feet long would be torn out by the
angular rocks, which caused considerable delay and undoubtedly entailed the loss of some gold. This unsatisfactory riffle
was replaced by a system in which the transverse and longitudinal types of riffles are combined.

The transverse type of riffle is considered to be the best gold
saver, being normal to the direction of flow of the material,
but at the same time offering considerable resistance to the
flow of gravel over it. The pole or longitudinal type of riffle
permits a free flow because the gravel and rocks pack between
the poles and permit the escape of some of the finer particles
of gold. To utilize the best features of both types of riffles,
3 x 4-inch timber crossties are placed in the bottom of the sluice
at intervals of four to six feet. Upon them are laid seven parallel lines of 25-pound rails, which are fastened into the sluice
with wooden blocks. The tops of the blocks are placed level
with the tops of the rails. These blocks serve the purpose of
riffles, without greatly retarding the flow of material through
the sluice. Along the upper 3,000 feet of the sluice the rails
are placed with the flat side down, but they are inverted in the
lower part to increase the sliding motion of the boulders.

The crossties beneath the rails act as cross riffles. As the
rails are held above the bottom by the crossties, a certain
amount of vibration is set up by the rocks that pass over them,
which causes the gold to settle to the bottom. At intervals of
about 100 feet cast-iron blocks that reach to the top of the rails
are fastened to 3 x 10-inch cross timbers. These blocks divide
the sluice into sections, so that in case a rail is dislodged only
one section will be affected and repairs can be made quickly.
However, since this type of riffle has been used no such trouble
has been experienced.

One advantage of the rail riffle is that the greater part of the
wear occurs at the center of the sluice. About 150,000 cubic
yards of material can be hydraulicked before the center rails
show any appreciable wear. When they become worn they are
moved to the outside position, and the outside rails are moved
to the center. In this way all the rails wear out at the same
time.

At Round Mountain, the handling of rocks and boulders too
big to be moved by the giants has been a problem. Various
methods have been tried; by one method the large rocks were

blockholed and blasted and the smaller ones broken by hand sledges. An attempt was made to run all the rocks through the sluice except the larger ones, which were broken by blasting, but the proportion of large rocks soon became too great and this method had to be abandoned.

The present practice is to blockhole the larger rocks and remove them with the boulders that can be handled conveniently

Figure 50. Car for handling boulders at
Round Mountain placer.

in either of two ways—(1) by rock cars that are hoisted out of the placer pit on a trestle and automatically dumped, or (2) by boom derrick. In the first method specially constructed flat-top cars built on standard 30-inch-gage mine trucks are used. The cars are wheeled from the pit on 12-pound rails, which are connected by switches to the rails leading to the trestle track. Figure 50 shows a loaded car ready to be hauled out of the pit. In using the boom derrick, a flat, heavily constructed timber platform, reinforced with iron, is employed. Chains are fastened to each corner of the platform to form a sling. A ¾-inch cable fastened to the sling passes to the derrick and electric hoist. The pole of the derrick is set at a slight angle from the vertical, so that the pole automatically swings out of the pit. A man is stationed at the dump pile to unhook two of the chains; then, by starting the hoist, the platform is emptied of the rocks upon it. Three platforms are used, so that two can be loaded while the third is in transit.

Standard jackhammer machines are used to blockhole the larger boulders. The depth of the holes drilled varies with the

Figure 51. Hydraulic giant in operation at Round Mountain placer.

size of the boulders and ranges from 6 to 12 inches. The explosive charge is from ¼ to 1½ sticks of 40 percent gelatine dynamite per hole.

The amount of water available at the placer banks is approximately 400 miner's inches (40 miner's inches are equivalent to one cubic foot per second). Two small giants or one large giant are used in each of the two pits, and are operated at the same time in order to permit cleaning up the sluices in one pit while hydraulic operations are going on in the other. Figure 51 shows a 7-inch giant in operation.

The amount of material that can be moved with 400 inches of water under a head of 350 feet is from 32 to 110 cubic yards per hour, depending on the height of the bank and the amount of rocks and cemented material.

The giants are not placed in the usual position, directly oppo-site the bank to be hydraulicked, but are set at an angle to the banks, because the angular rocks do not pass readily into the sluice and the force of the water is needed behind the material rather than in front of it.   It is also necessary for the giant operators to exercise care not to move too much soil without a proper proportion of rocks; otherwise a pile of rocks will accu-mulate, and, lacking water of the proper density and buoyancy, an undue proportion of rocks will have to be removed by the methods previously described.   Also, care must be used to keep the sluice filled nearly to capacity with gravel and water, because if the water is allowed to get too low the gravel will settle and choke the sluiceway.

At times a considerable amount of cemented material in the placer banks makes blasting necessary.   Holes five inches in diameter are drilled with a Star drilling rig.   These holes are drilled to bedrock, about 50 feet apart, parallel to and 25 feet from the edge of the bank.   One method of doing this is described: Four holes were drilled to an average depth of 46 feet and each hole was sprung with 15 pounds of 40 percent gela-tine dynamite.   After being sprung, each hole was loaded with 175 pounds of Judson powder and the charges detonated.   The blast thoroughly shook and loosened 12,000 cubic yards of mate-rial, or about 17 yards per pound of powder.   The cost of this work, including setting up the drill, moving it from hole to hole, sharpening drill bits, cost of explosives and labor, was $582.95, or about 4½ cents per cubic yard of material.

The rough bedrock is cleaned as well as possible by the moni-tors, after which it is allowed to dry, and is then cleaned by hand, specially constructed small tools and whisk brooms being used to remove the gold from the crevices.   The cost of cleaning bedrock, per cubic yard of total material moved, varies with the height of the bank, character of bedrock and several other fac-tors.   In one year, when 128,000 cubic yards of gravel were sluiced, the cost of cleaning bedrock was 15½ percent of the total cost of the year's operations.   During this same year 72.7 percent of the gold was recovered in semimonthly clean-ups, 11 percent in bedrock cleaning, 9 percent from the lower reaches of the sluice, and 7.3 percent in milling residual material in the sluice.

At the end of each successful season the sluice is cleaned up for a distance of several hundred feet beyond the point to which the semimonthly clean-ups extend; and in the fall, at the close

of the placer season, all rails are removed from the sluice, and
the material remaining in the sluice is allowed to dry, after
which it is hauled to the mill and treated the same as ore. This
material has a value of about $60 per ton.

Quicksilver is added from time to time while sluicing is in
progress, the amount depending upon the yardage and the value
per cubic yard as determined by panning tests. Clean-ups are
made semimonthly by running down about 200 feet of the upper
end of the sluice, removing the lining boards, blocks, rails, and
crossties while a small amount of water flows through the sluice.
The blocks, rails, and crossties are scrubbed thoroughly with
brooms to remove any particles of adhering gold or amalgam.

Figure 52.   Placer treatment plant, Nevada Porphyry Gold Mines, Inc.,
Round Mountain District.

More quicksilver is then added and the flow of water increased,
so that the material will move slowly down the sluice. As this
goes on, the rocks are forked out with eight-tine sluice forks.
The gold or amalgam moves down behind the lighter material
and is scooped up, placed in buckets, and taken to the mill for
grinding in a clean-up pan, as much of the gold is associated
with quartz and other material. The gold is then amalgamated,
retorted, and melted. It is then sent by registered mail to the
American Smelting and Refining Company at Selby, Calif. The
bullion is insured for its full value from the time it is deposited
in the post office at Round Mountain. This method has proved
cheaper than transporting to the railroad and shipping by express.

Yardage measurements are made twice a month while operations are in progress. Preparatory to mining, the placer banks

are drilled and sampled. In sampling, an area is surveyed and divided into blocks 100 feet square. Holes are drilled or pits sunk at each corner of a square. The drill holes, or test pits, are sampled at intervals of 10 feet. Each sample contains from 60 to 100 pounds of material. The samples are dried thoroughly, weighed, and passed through a small sluice box about 15 feet long and 6 inches wide, having Hungarian-type riffles and containing quicksilver. After each sample is sluiced, the box is cleaned up, the recovered amalgam retorted, and the amount of gold in proportion to the weight of the sample determined.

In computing yardage, each square is divided into four equilateral triangles and a separate computation made for each tri-

Figure 53. Mining placer gravel with electric power shovel, Nevada Porphyry Gold Mines, Inc., Round Mountain District.

angle. This method is more accurate than using the four holes at the corners of the 100-foot squares.

In order to work placer ground below the hydraulic grade line, mechanical placer equipment has been installed. The plant was completed in November, 1934. It has a capacity of 1,000 cubic yards per 24 hours. A general view of the treatment plant is shown in figure 52.

The gravel is mined with a ¾-cubic-yard-capacity Bay City electric power shovel mounted on caterpillar treads. In figure 53 the shovel is shown working against a 45-foot placer bank.

The gravel mined by the shovel is dumped into the 1½-cubic-yard hopper of a portable belt conveyer that is 50 feet long and 30 inches wide. The portable conveyer feeds onto a main conveyer belt which is 550 feet long, 30 inches wide, and inclined

11 degrees. The main conveyer discharges into the trommel at the washing plant.

The trommel is 6 feet in diameter, 28 feet long, and makes 9 revolutions per minute. The slope of the trommel is $\frac{1}{3}$ inch per foot. One-half the length of the trommel is used as a disintegrating section. This section is covered with sheet iron and lined with 1 x 4-inch strap iron. To assist in breaking up the lumps of cemented gravel, six $3\frac{1}{2}$ x $3\frac{1}{2}$ x 1-inch angle irons are riveted to the inside. The screening section is made of plate punched with $\frac{1}{2}$-inch holes.

The trommel oversize is discharged to a tailings stacker that can be swung in an arc of 90 degrees. An auxiliary horizontal

Figure 54.   Gravel-washing plant at Nevada Porphyry Gold Mines, Inc., Round Mountain District.

stacker belt takes the gravel from the main stacker and carries it to the waste pile in a nearby gulch. The auxiliary stacker is 50 feet long, 24 inches wide, and is mounted on trucks running on light track, so that it can be moved sideways.

The trommel undersize is distributed to three riffle boxes each 3 feet wide and 50 feet long. One-and-one-half-inch angle irons are used for riffles.

From the riffle deck the tailings discharge by gravity to a home-made drag classifier for dewatering. A view of the classifier and other placer equipment is shown in figure 54. The classifier is equipped with $2\frac{1}{2}$ x $3\frac{1}{2}$-inch angle irons bolted to a belt 48 inches wide. Angles are spaced one foot apart. The head pulley of the classifier is two feet and tail pulley eight feet in diameter. The classifier tailings are discharged onto the stacker belt with the oversize material from the trommel.

The classifier overflow is carried by launder and ditch to a series of three settling ponds. After settling, the water is returned to the placer plant by two-stage Byron-Jackson pump. About three-fourths of the water is added to the trommel and the other one-fourth to the sluices. Water is fed to the disintegrating section of the trommel through two 2-inch pipes with open ends, and to the screening section through a 4-inch perforated pipe.

In places the gravel is cemented so that blasting is necessary. Holes are driven into the bank to a depth of 15 to 20 feet with a ½-inch pipe and compressed air. A water jet was tried, but the compressed air jet proved to be more satisfactory. After the holes are sprung, each hole is loaded with about 25 pounds of Hercomite powder and detonated.

In mining, the boulders are cast to one side by the power shovel. Boulders too large to be handled by the shovel are blockholed with jackhammers and blasted.

Clean-ups are made about every two weeks. Quicksilver is put into the sluices about once each shift, the amount depending on the richness of the gravel. Grab samples are taken off the main conveyer belt every hour and panned. The tailings are panned from time to time.

Electric power is purchased from the Nevada-California Power Company. Distribution of power at the placer plant according to motor rating is as follows:

	HORSEPOWER
Power shovel	50
Portable conveyer	5
Main conveyer	30
Trommel	30
Drag classifier	10
Main stacker	15
Auxiliary stacker	5
Water return pump	25
Total	170

When operating on a 3-shift basis, 5 men are required to run the plant on each shift, with two additional men on the day shift for general work. The cost per cubic yard of gravel treated is between 25 and 30 cents when plant operates to capacity of 1,000 cubic yards per day.

The Union District, also known as the Berlin or Ione District, is in northwestern Nye County on the west slope of the Shoshone Range, about 46 miles southwest of Austin, Nevada. The District was discovered in 1863, and the town of Ione was established and made the county seat of Nye County in 1864. The amount of placer gold mined in the District has been small.

In 1932 an attempt was made to work a placer deposit one mile southwest of Ione. The equipment used for mining and treating the placer material consisted of two dragline scrapers, a specially designed gold-saving machine, and a power plant. The attempt was unsuccessful and was abandoned after a short time. A view of the plant is shown in figure 55.

The placer material consists of surface debris, sand, and soil,

Figure 55. Placer gold machine of special design at Ione.

which covers a gentle slope to a depth of several feet. Water for washing the placer material was piped by gravity from Ione.

In 1934 and 1935 small quantities of placer gold were obtained in the District by dry washing.

PERSHING COUNTY
PLACERITES DISTRICT

The Placerites District is about eight miles south of Scossa and 47 miles north of Lovelock in northeast Pershing County. The placers occur in low hills in the vicinity of T. 32 N., R. 29 E. According to E. J. Quirk of Rosebud, the first placer mining in the District was done in the early seventies by a man called "Mahogany Jack," and his three partners, who took out about

$30,000 in placer gold by hand methods. In the nineties some placering was done, the gravel being hauled to Rabbit Hole Springs and worked in rockers. In 1928 C. J. and E. G. Stratton worked eight months and recovered $5,000. The gravel contained from 25 cents to $5 per cubic yard.

In 1929 a stock company, called the Nevmont Placer Mining Company, was organized and acquired control of 4,000 acres of potential placer mining ground. Equipment was installed to work the placers on a large scale. A reservoir about 1,000 feet long, 35 feet wide, and 25 feet deep was excavated in the alluvium near the foot of the hills. (See figure 56.) The water rose in the reservoir to within 20 feet of the rim, and it was thought that the water supply would be sufficient for profitable large-scale operations. From the reservoir the water was

Figure 56. Water reservoir and pumping plant for placer operations, Placerites District.

pumped to several tanks on a hill above the placer area. The combined capacity of the tanks was 145,000 gallons. A dragline scraper was installed to mine the gravel. The gravel was screened in a trommel and the undersize was sluiced. The amount of water available in the reservoir was inadequate, and in 1932 a 5-inch gravity pipe line was laid from Cow Springs to the ground, a distance of eight miles. In addition, the dragline scraper was replaced by a gasoline power shovel, but owing largely to the inadequate water supply, the venture was not commercially successful. In 1931 the company for a

time treated an average of 140 cubic yards per day, but the
gold output was not reported.

There are at least five gulches in the District that are known
to have carried gold values. No doubt the best spots were worked
intensively in the early days. The early-day diggings have been
largely obscured by the effects of subsequent cloudbursts.

In 1934 and 1935 a little placer mining was done by "snipers"
who worked with dry washers at the heads of the shallow ravines.
The gravel worked by these small operators was from 18 inches
to 6 feet deep. The bedrock in the greater part of the placer
area is composed of sedimentary material, principally slate and
shale. Most of the gravel is small, although some large boulders are present. The gold is coarse and presumably has not
traveled far. Its fineness varies from 730 to 900. A large
amount of black sand is concentrated with the gold in working with dry washers.

RABBIT HOLE DISTRICT

The Rabbit Hole District is north and west of Rabbit Hole
Spring and about five miles northeast of Placerites. The Rabbit
Hole District adjoins the Rosebud District, and it may be con-
sidered part of the latter.

According to E. J. Quirk a placer location was made in the
Rabbit Hole District as long ago as 1900, but no work was done
in the area until 1916, when the Wogan brothers located several
claims and took out about $3,000 in placer gold. Mr. Quirk
worked during several summers following the year 1916 and
recovered $600 by panning, water being hauled from a well in
Rosebud Canyon. After a cloudburst, Mr. Quirk has picked up
small nuggets of gold from the surface in Coarse Gold Canyon.

No accurate figures are available on the total placer production
of the District. In 1935 this was one of the important placer
districts in the State. During the summer of 1935 as many as
150 men worked in the District at one time with dry washers
and rockers. The average returns from this work netted the
operators wages.

Placer gold has been found in a number of gulches and ravines
tributary to Rosebud Canyon. The best values are found near
the heads of the ravines and on the slopes above the ravines.
The area of the placer ground is about five square miles.

The depth of the gravel worked in Coarse Gold Canyon, where
most of the placer mining has been done, varies from 2 to 12
feet, averaging about 4 feet. Pay gravel lies on a false clay

bedrock, and beneath this false bedrock the gold values diminish. Several shafts have been sunk to depths of 68 to 100 feet in the canyon in an endeavor to reach true bedrock, but the flow of the water encountered was too great to be overcome by the simple means available to the small operators who did the work. The surface gravel is rough and angular and consists largely of minus 1-inch material. A few boulders are present, but do not exceed 14 inches in diameter.

Most of the gold recovered is nuggety. The gold from the District is easily recognizable as the nuggets are all flat. Nuggets range in value from 20 cents to several dollars. The largest nugget found in the District in 1935 weighed 9 pennyweight and 6 grains. The distribution of the gold is very erratic; rich pockets have been found that average as high as $50 per cubic yard. Its fineness averages about 900. A considerable amount of black sand is present.

In recent years several attempts have been made to work the placers on a large scale. In the fall of 1932 a company installed a washing plant at Rabbit Hole Springs. A graded road, three and one-half miles in length, was built from the placer to the washing plant. A fleet of five motor trucks conveyed the gravel to the plant. The plant set-up and treatment was as follows: Gravel was dumped from the trucks into a hopper, from which it was fed to a Barber Green belt conveyer, which delivered it to a trommel 12 feet long and 4 feet in diameter, covered with ¼-inch mesh screen. Fifty gallons of water per minute were required to wash the gravel through the trommel. The trommel oversize was discharged to waste and the undersize was washed in two power-driven sluice boxes, one of which was 16 feet long and 4 feet wide and the other 12 feet long and 3 feet wide, each set on a grade of 3 inches per foot and fitted with transverse riffles. Below each rocking box were short stationary sluice boxes 16 feet long and 19 inches wide set on a grade of 1½ inches per foot in the bottom of which were amalgamated copper plates extending the full length of the boxes. Above the upper copper plate a 16-mesh screen was used to separate and carry away the coarse sand. After passing over the screen, the coarse sand was washed out of the box over a steel plate placed above the lower amalgamated plate. Amalgam traps were employed at the lower end of the sluice boxes. Water was supplied by two centrifugal pumps at the rate of 50 gallons per minute. Power for the plant was developed by a 28-hp. gasoline engine. The owners stated that the initial operation in October, 1932, yielded an average of $2 per cubic yard, and that the capacity of the plant was greater than 100 yards per day. Operations were suspended in 1935.

An unsuccessful attempt was made to work the surface gravel in the District with a power shovel and a battery of ten dry-washing machines connected to a single power unit. The gas shovel employed had a capacity of three-eighths cubic yard. The treatment plant was moved ahead with the power shovel.

In November, 1935, about 50 men were engaged in placering in the District, using dry washers and rockers.

A power dry washer used by T. J. Basford is shown in figure 57. The dry washer is the ordinary bellows type. The

Figure 57. Dry washer driven by gasoline engine, Rabbit Hole District.

power unit consists of a Maytag washing - machine engine mounted at one side of the dry washer in order to escape the dust. The drive shaft is made of 1-inch pipe bolted to the center of the pulleys. With two men shoveling, the capacity of the machine is about 12 cubic yards per day. In 1935 Mr. Basford worked six months, and recovered $1,000 in gold with the machine shown. Most of the gravel is fairly dry, so that a good recovery can be made with dry washers. However, some of the fine gold is locked in clayey material. This is demonstrated by the fact that the tailings from dry washing operations can be re-treated, after being exposed to the weather for some time, and some gold recovered.

The cost of the machine complete is $92; the Maytag engine

cost $38, the dry washer $50, and the belts and pulleys $5. The engine is rated at ¼ horsepower and burns one quart of gasoline in six hours.

Another type of power dry washer used in the District is shown in figure 58. This machine was made by C. B. Richardson of Sulphur, Nevada. The power unit consists of a Maytag engine that is belt-connected to a generator, which is made from a converted ½-hp., 110-volt, 1,725-r.p.m. motor. The motor is mounted in a dust-proof housing on the rear of the dry washer. The rating of the motor is ¼ horsepower. The motor is connected to the generator by about 50 feet of wire cable. The advantage of this type of machine is that the dry washer can be moved from place to place within the length of

Figure 58. Electric-driven dry washer, Rabbit Hole District.

the cable without moving the power unit. The capacity of this machine, with steady shoveling of two men, is from 12 to 15 cubic yards per eight hours. The engine operates on three fourths of a gallon of gasoline per eight hours and one pint of oil per week. The cost of the machine is as follows:

Maytag engine	$42.50
Dry washer	50.00
Generator (used)	20.00
Motor (used)	5.00
Cable, belts, pulleys	3.00
Total	$120.50

Several operators in the District were using Power-lite outfits to run their dry washers. Space does not permit a description of the great variety of dry washers in use in the District. Each type of machine had its adherents. In many cases the machines differed only in design, the principle of operation being the same. It was generally conceded, however, that the power-driven dry washer was far superior to the hand-driven machine.

At Barrel Springs, in the canyon of the same name west of Coarse Gold Canyon, Otto Jancke and three partners have operated a placer plant for several years, treating gravel from their own ground. Over 3,000 acres of placer ground is held by Jancke and his associates. The gravel is hauled three miles to the spring, where sufficient water for placer purposes has been developed by sinking a well in the canyon. The gravel is treated in a power rocker driven by a Chevrolet automobile engine. The pump at the well is operated by the same engine. The capacity of the plant, with four men, is about 16 yards per day. In 1935 three months' work netted each of the four men $1,000. In 1934 each man earned $2,200 in about eight months. Most of the values recovered are in the form of small nuggets worth from 50 cents to $3. The largest nugget found had a value of $17.50. Gold has a fineness of 895.

ROCHESTER DISTRICT

The Rochester District lies in the Humboldt Range in central Pershing County, about 22 miles northeast of Lovelock, Nevada. The placer deposits were discovered in the early sixties by men from Rochester, N. Y. The District attained considerable prominence when important silver veins were discovered by Joseph Nenzel in 1911.

The placers have been worked intermittently for many years, and according to report, it was the practice for a number of prospectors to procure annual "grub stakes" from the placer deposits in Rochester Canyon. The early placer workings have been covered with large accumulations of cyanide mill tailings. The auriferous gravels are from 50 to 100 feet thick. The bedrock is quartzite and the values are said to occur in the gravel from surface to bedrock. The gold is rough and crystalline, with a fineness of about 800. Much black sand is associated with the gold.

In 1931 placer gold was discovered in Gold Springs Gulch near the mouth of Rochester Canyon. Here the placer deposits consist of loosely cemented gravel lying on a false bedrock of clayey volcanic ash. In 1931 and 1932 the gravel deposits were prospected and a small amount of gold recovered by dry washing.

The pay gravel is from four to eight feet deep. In recent years there has been a small amount of placer mining in this area.

In Limerick Canyon, about one mile north of Rochester Canyon, placer gold deposits were worked in former years. This canyon is about six miles long and rather narrow and no doubt it was worked intensively in the early days by hand methods. In 1914 placer gold was discovered in the east end of Limerick Basin, where gravel occurs in a depression about one mile long and one and one-half miles wide. This location was profitably worked by small - scale methods. In one place where the gravel averaged about six feet in depth, two men are reported to have taken out $60 to $100 per day with a rocker. The pay gravel was from 12 to 15 inches thick in an old channel above bedrock. The pay streak was covered in places with from four to six feet of overburden.

Near the west end of Limerick Basin, Andrew B. Puett has worked his placer claims, totaling 146 acres, for a number of years. Here the gravel is from 2 to 38 feet deep, averaging about 11 feet. Bedrock is andesite and schist. The placer material is composed chiefly of small angular rock fragments, sand, and soil. Angular gold, both coarse and fine, is distributed throughout the gravels from surface to bedrock. Fineness of the gold is from 810 to 880. The largest nugget found by Puett weighed 10 pennyweight, 8 grains. A large amount of black and brown sand concentrate is associated with the gold. The gravel was hauled to McCarthy Springs, one and one-quarter miles below the property, and the gold recovered in a rocker.

In 1931 the Limerick Canyon Mining Company erected a small plant on the Puett property. The plant had a capacity of six cubic yards per hour. Water for placering was piped from McCarthy Springs through a 2-inch pipe line. The gravel was first handled by dragline scraper and later by a small power shovel. The plant was operated only a short time. It is reported that operations were unsuccessful because of water shortage.

In 1935 several small operators placer mined in Limerick Canyon for short periods.

From 1914 to 1917, inclusive, W. W. Walker and E. T. Walker of Lovelock sank a series of shafts from 15 to 20 feet deep near the upper end of Limerick Basin and drift mined over 2,000 feet of channel. Pay gravel on bedrock was reported to have averaged $35 per cubic yard. Water being scarce in the vicinity, the gravel was either hauled six miles to a spring, or the water was hauled to the diggings. Over $40 per day was recovered with a rocker in one 60-day season, and a few rich patches of gravel were found that gave $5 per pan. From 1914 to 1917 the mine netted the owners $18,000.

As pointed out by Schrader,[20] the Limerick Canyon placers are of special geological interest for they may indicate the course of an ancient stream channel that may have crossed the range by way of Spring Valley Pass.

In Walker Canyon and American Canyon, on the east side of the range, well-washed gravels containing granite pebbles occur, although granite is not present in the neighboring rocks. In Limerick Canyon and Basin the gravels are a heterogeneous mixture of rough and subangular rock fragments and well-rounded pebbles, the latter probably having been derived from the erosion of the ancient stream channel near the crest of the range.

ROSEBUD DISTRICT

The Rosebud placer area is in northern Pershing County, about 12 miles south of Sulphur, Nevada. According to E. J. Quirk, some of the ravines tributary to Rosebud Canyon were worked in the seventies by Chinese placer miners, and several thousand dollars in gold was recovered. The placers have been worked by the Delavaga brothers in recent years. These operators state that they have recovered about $3,000 in gold by their work.

About six shafts have been sunk to bedrock in Rosebud Canyon. The shafts are from 40 to 90 feet deep. Some gold was found while sinking these shafts. Several shafts encountered a considerable flow of water near bedrock.

About 1930 a stock company, called the Associated Royalty Mining Company, was organized and 4,160 acres of potential placer ground was acquired. In recent years the company leased a portion of their ground to small operators, who recovered some gold with rockers.

SACRAMENTO CANYON

Sacramento Canyon is in the northwest flank of the Humboldt Range in central Pershing County, about five miles east of Oreana, a station on the Southern Pacific Railroad.

Placer gold was discovered in the canyon in 1912 and a small amount of prospecting was done. There has been no placer mining in this area in recent years.

[20] Schrader, F. C., The Rochester Mining District, Nevada: U. S. Geol. Survey Bull. 580, 1913, pp. 371, 372.

SAWTOOTH DISTRICT

The Sawtooth District is in northern Pershing County near the Humboldt County boundary line. It is about 50 miles north of Lovelock and 25 miles east of Sulphur, Nevada, a station on the Western Pacific Railroad. Placer gold was discovered in the District in 1931 by Rufus Stevens as a result of prospecting stimulated by the discovery of the gold veins at Scossa, which is about 12 miles south. Shortly after the discovery many claims were staked, and before the end of the summer of 1931 as many as 35 men were dry washing in the District with encouraging results. It is reported that some individuals recovered as much as $35 per day. Dry washing continued in 1932, and, in addition, the Oregon-Nevada Mining Company was organized to work the gravels on a larger scale. J. R. McCroden of Lovelock states that in 1932 he worked in the District with a hand rocker, and an average of 12 wheelbarrow loads of gravel netted him $7. Water for rocking was hauled from Mandalay Springs, three miles distant.

Placer gold has been found over a fairly level area of about six square miles. An unusual feature of the placer is that the best values are found at shallow depths. Much of the gold has been found above a false clay bedrock at depths of eight inches to two feet. No shafts have been sunk in the District to prospect the true bedrock. The gravel is rough and angular, with a small percentage of boulders. A considerable amount of clay is present in places, and the clayey material has to be dried and pulverized before a satisfactory saving of gold can be made with dry washers. Some barren white quartz fragments are found on the surface and prospectors use the quartz as an indicator of gold concentration. The gold is coarse and rough and averages about 880 fine. Small nuggets worth up to $4.50 have been found.

The best placer ground in the District was owned by the Oregon-Nevada Mining Company. This company used a caterpillar tractor and scraper to haul the top gravel to a belt conveyer, which discharged into a trommel screen with ¾ - inch holes. The coarse was discharged to waste and the undersize went to a disintegrating machine equipped with wooden paddles. After disintegrating, the material was treated on two home-made concentrating tables. Concentrates obtained from the tables were cleaned with a hand rocker. The plant was closed down in 1934, and the property acquired by others.

Water was obtained from Mandalay Springs, Mandalay mine shaft, and a well on the property. It is said that the water was so slimy after it was used that it could not be reclaimed.

An attempt was made to treat the gravel on a large scale by dry washing, but this venture was not commercially successful.

In 1935 placer activity in the District was confined to small dry-washing operations.

SIERRA DISTRICT

The Sierra District, also known as the Dun Glen, Oro Fino, Sunshine, or Chaffee District, is in the northern part of the East Range, about 40 miles northeast of Lovelock and 12 miles east of Mill City, the latter a station on the Southern Pacific Railroad.

The first lode discovery in the District was made in 1863 and the placers a short time after. The placer deposits occur over a large area and include deposits in Auburn, Barber, Wright, Rock Hill, Dun Glen, and Spaulding Canyons. The amount of placer gold mined is estimated to be about $4,000,000, principally from Barber, Wright, Auburn, and Rock Hill Canyons. The bulk of it was produced by Chinese placer miners during the eighties and nineties.

Some exploratory work was done in Dun Glen Canyon during 1932 and 1933 by tunneling and sinking shafts, but no equipment for treating the placers was installed. The Dun Glen placer has a length of about 4½ miles, with a maximum width of 800 feet. The average width is about 200 feet. In past years a number of shafts have been sunk along the course of the canyon for testing purposes. The depth of these shafts is from 18 to 40 feet to bedrock. It is reported that from five to six feet of gravel lying above bedrock will average between $2 and $3 per cubic yard at the old price of $20.67 per ounce for gold. Drift mining the gravels has been hampered by the large flow of water encountered near bedrock.

In 1931 a company organized in Los Angeles acquired control of 1,000 acres of placer ground. The company installed a specially designed land dredge having a capacity of 100 cubic yards per day. Gravel was excavated by power shovel. The company ceased operations after only a short time.

In 1934 sluicing was done on a small scale in Barber Canyon by Barney Sorenson of Salt Lake City. The placer ground was worked by first stripping the overburden to a depth of 12 feet with a caterpillar tractor and scraper and then hydraulicking the

gravel near bedrock. A canvas hose and ordinary fire-hose nozzle were employed to sluice the gravel to the washing plant. The
site of the hydraulic operation is shown in figure 59. The washing plant, in the canyon below the placer ground, is shown in
figure 60. This washing plant consisted of a combination disintegrating and screening trommel 10 feet long and 4 feet in
diameter. The disintegrating section of the trommel was six feet
in length and the screening section four feet, the latter being
equipped with 1-inch holes. The trommel was belt driven by an
old automobile engine. The sluice box below the trommel was
36 feet long and had Hungarian riffles. The plant is said to
have had a maximum capacity of 10 cubic yards per hour. Water

Figure 59. Hydraulic operation in Barber Canyon.
(Picture courtesy of Mr. Thomas Varley.)

for placering was obtained from a well in the canyon above the
plant. This well is 105 feet deep and the water level is within
40 feet from the surface. Water pressure for hydraulicking was
furnished by a 4-inch centrifugal pump driven by a gasoline
engine.

The canyon is about two miles long and has an average width
of 300 feet. The average depth to bedrock is 30 feet. No large
boulders are present in the gravels. The best values are on bedrock and on benches at the sides of the canyon. Gold is coarse
and averages 875 fine. A small amount of black sand is associated with the gold. The average value of ground is reported

to be 28 cents per cubic yard, including the low-grade surface material.

In 1935 the hydraulic equipment was replaced by a patented placer machine called a "desert sluice," invented by G. W. Rathjens, chief engineer, United States Smelting and Refining Company. A 1-cubic-yard gasoline shovel was used for mining the gravel. This placer machine is said to operate successfully with

Figure 60. Washing plant for placer gravel, Barber Canyon. (Picture courtesy of Mr. Thomas Varley.)

less than 1 percent of the water ordinarily used in placer operations and to have a capacity of 15 to 25 cubic yards per hour. A description of the machine is given in the Engineering and Mining Journal, issue of August, 1935.

In Spaulding Canyon, in 1932, about six groups of men were employed in small-scale placer mining along a two-mile length of the canyon. These groups are reported to have made wages.

SOUTH AMERICAN CANYON

South American Canyon is about one mile south of American Canyon in the Rochester Mining District. This canyon was worked by Chinese for many years; the last survivor of the former colony of Chinese placer miners died in about 1928. Judging from the extent of the placer excavations in the canyon, a large amount of gold was recovered. About half-way up the canyon from Buena Vista Valley is a bowl-shaped depression about 1,500 feet wide, which formed a natural concentration area for placer gold. This basin is scarred by numerous shafts and tailings piles. None of the shafts are over 15 feet deep. The gravels are subangular and contain no large boulders. The gold was recovered by the Chinese with rockers and by panning. Water for this work was obtained from several springs in the lower part of the canyon. The best ground in the canyon undoubtedly has been worked by the Chinese, and recent attempts to find pay gravels have not been encouraging.

SPRING VALLEY DISTRICT

The Spring Valley District is on the east flank of the Humboldt Range in central Pershing County, about 30 miles northeast of Lovelock. Although lode mines in the District were active as far back as 1868, placer mining did not begin until 1881. The Spring Valley and American Canyon placer areas have attained the largest output of placer gold in the State. Ransome[21] states:

The placers were first worked by Americans, who are reported to have taken out gold to the value of $1,000,000. The ground, however, soon passed into the possession of the Chinese, who formed a considerable settlement in American Canyon and mined the gravels with skill and assiduity by drifting from countless narrow shafts ranging from 40 to 85 feet deep. How much gold they obtained is unknown, but some estimates, doubtless much exaggerated, place the total at about $10,000,000.

Old-time residents of Lovelock informed the writer that between the years 1884 and 1895 about 3,000 Chinese placer miners were employed in the District. According to Locke,[22] in the early nineties hundreds of Chinese were working the gravels in American Canyon. Ample evidence of the intense activity that prevailed remains in the hundreds of shallow shafts and piles of gravel spread over two miles of the canyon. The Americans leased the ground to a Chinese placer operator on a royalty basis. This operator brought in hundreds of his

---

[21] Ransome, F. L., Notes on Some Mining Districts in Humboldt County, Nevada: U. S. Geol. Survey Bull. 414, 1909, p. 12.
[22] Locke, Ernest G., Reawakening of an Old Placer Camp: Min. & Sci. Press, vol. 107, 1913, p. 373.

fellow countrymen, to whom he subleased the ground in blocks 20 feet square. Each lessee sank a shaft to the pay streak and mined out the pay gravel within the limits of his lease. It is stated that each block of 400 square feet produced an average of $1,500 to $3,000 in gold dust and nuggets. As the Chinese were secretive as to the amounts of gold taken out, and as the bulk of this gold never found its way to the mints in this country, the aggregate amount of gold recovered will never be known accurately.

The source of the gold in American Canyon has not been definitely traced, but it is believed to have originated in a range of porphyry hills through which the canyon was cut by a stream. In working the gravels downstream, the bedrock was found to be porphyry for a distance and then limestone. The pay streak was found on a false bedrock of clay at an average of 60 feet from the surface, shallower at the upper end of the diggings, and deeper at the lower end. A shaft was sunk to limestone bedrock at the lower part of the diggings, but the pay streak still retained its position at about 70 feet from the surface, and little or no gold was found below the false bedrock and none on the limestone.

Spring Valley, in addition to the neighboring Dry Gulch and American Canyon, was also the scene of small-scale placer mining by Chinese in the early days. Spring Valley Canyon is about four miles long and averages about 1,000 feet wide. Judging from the numerous piles of boulders and cobblestones the canyon has been pretty thoroughly worked.

In 1910 the Federal Mines Company installed a dredge to work the lower part of the canyon. The dredge worked in stream and bench gravels that sampled from 22 to 30 cents per cubic yard, and varied in depth from 20 to 30 feet. Gold was present from the surface down in a series of gravel layers underlain by seams of clay. The alluvium contained many large boulders, which interfered considerably with dredging operations. The gold was coarse, and nuggets worth from $3 to $5 were common. The gold had a fineness of about 700.

One operating disadvantage in dredging was the scarcity of water, which had to be piped from Indian Creek, five miles distant. In addition, considerable difficulty was had in keeping the dredge afloat because of the steepness of the canyon and the old underground workings, which threatened to drain the dredge pond.

According to the old records of the company, the output of

the dredge from October 15, 1912, to January 17, 1914, was
$10,902.43. During this period 58,494 cubic yards of gravel
were dredged with an average recoverable value of 18.6 cents
per cubic yard.

A view of the remains of the dredge on the site of operations
is shown in figure 61. The dredge was made of wood. The
digging ladder was of the open connected - link type, which is
best adapted to ground containing large boulders. The capacity
of the dredge was 2,000 cubic yards per day. After screening
and washing in a trommel, the undersize passed on to tables
that sloped 1 inch in 15 inches and which were equipped with
angle-iron riffles. There were 40 linear feet of tables 6 feet in

Figure 61. Remains of old placer dredge in Spring Valley. Chinese placer
diggings partly obscured by sagebrush in foreground.

width, and, in addition, the gold-saving equipment consisted of
32 feet of sluices and riffles.

The stacker consisted of a continuous chain of buckets closely
connected and driven by rope transmission from the main drive.
The power for operating the dredge was furnished by four
distillate-burning engines having a sea-level rating of 100, 55,
30, and 8 horsepower, respectively. The altitude at which the
dredge operated was 5,000 feet, so that the total effective horsepower was only about 81 percent of the sea-level rating, or 156
horsepower. The 8-hp. engine drove the dynamo for lighting
and a small pump for priming; the 30 - hp. engine drove the
winch; the 55-hp. engine operated the double 10-inch centrifugal pump; and the 100-hp. engine ran the digging ladder, screen,

6

and stacker. Head and side lines were used instead of spuds for holding the dredge in position.

W. H. Harris, now living in Winnemucca, Nevada, informed the writer that he worked a virgin section of placer ground at the lower end of Spring Valley from 1921 to 1924 and recovered a small fortune in placer gold with a rocker. He found, in sinking a shaft 40 feet, that the gold was concentrated at five different horizons on strata of clay. The lowest of the five pay streaks carried the best values.

In 1935 an attempt was made to work hillside placer material at the upper end of Spring Valley. A ¼-cubic-yard gasoline shovel was used to mine the gravel. The gravel was screened through a trommel with 1-inch holes and the undersize was treated by sluicing. Water for sluicing was obtained from a well sunk in the canyon. Flow of water in the canyon along bedrock is reported to be about 60 gallons per minute. This operation was not successful and lasted only a short time.

The only mining in the District in the fall of 1935 was done by a few lessees who were working the hillside gravels and recovering the gold with hand-operated dry washers. The hillside gravels are slightly cemented and contain a number of large boulders. Depth of gravels varies from 3 to 28 feet. Best values are found on the rough bedrock. The average daily earning per man was about $2.

UNIONVILLE DISTRICT

The Unionville or Buena Vista District is on the east slope of the Humboldt Range in central Pershing County, about 30 miles northeast of Lovelock, Nevada. The District was organized in 1861. During the early sixties the region in the vicinity of Star Peak was the scene of intense silver mining excitement, and one of the goals of the rush that took place at the time was Unionville. Among those drawn to the District in the general stampede was Mark Twain, who has left us a rich legacy in the colorful descriptions of early mining activity in that exuberant volume "Roughing It."

Placer gold was discovered early in Buena Vista Canyon, in which the town of Unionville is situated. Placer gold was found also in a neighboring canyon, which is reported to have been worked in the late forties, probably by the Spaniards.

Above the town of Unionville a new placer discovery was made in 1931. Here the gravel is from six to ten feet deep with a pay streak on bedrock from two to three feet thick. In 1932 several small-scale placer operations were carried on in the District, and the earnings per man were said to be as high as $10

per day. In 1935 there was a small amount of placer activity.
Operations are handicapped by scarcity of water.

In the early days some placer mining was done in Indian
Canyon, eight miles south of Unionville. The biennial report of
the State mineralogist for 1875 and 1876 states: [23]

The ravine north of the Eagle mine has been located for three miles
in length and four hundred feet in width. All the gravel in this ravine
contains more or less gold. In places, as much as two ounces of dust
to the man has been taken out in a day, and one streak of pay dirt gave
fifteen ounces to two men for one and one half days' work with one
rocker. The supply of water here is very limited, so that very extensive washings cannot be carried on. Rockers are used except for a
short time in the winter, when the supply of water is more plentiful,
and then the long-tom is operated. The gold obtained here is coarse
and worth about seventeen dollars per ounce. During the month of
September last (1875) about forty rockers were in use, and about thirty
white men and one hundred and twenty-five Chinese were at work. It
is impossible to get from the Chinese any information as to the amount
of gold extracted, but it is thought that they get about two dollars a
day to the man. In several of the ravines making out of the east side
of the range of mountains between the Eagle mine and Unionville,
good prospects of gold are found, but owing to the scarcity of water
these placers are not worked.

WEAVER CANYON

Weaver Canyon is one mile south of the mining town of
Rochester. The canyon is about two miles long, and takes a
westerly course. It has been prospected, and a small output
obtained by hand methods. The gravels are angular and contain some large boulders. The ores of the Colligan mine and
other mines at the head of the canyon were unusually rich in
gold and it is thought that the placer gold was derived from the
erosion of the veins in these mines.

WASHOE COUNTY
LITTLE VALLEY

Little Valley, in which profitable placers were found, lies in
the southwestern corner of Washoe County. The valley occupies
a narrow trough four miles long, having a north and south axis,
on the eastern slope of the Sierras at an altitude of 6,500 feet.
Although Little Valley adjoins the Voltaire District of Ormsby
County on the south, it is wholly in Washoe County and does
not constitute a part of any organized mining district. The
placer deposits were worked in the early days and the total
yield is reported to have been in the neighborhood of $100,000.
Evidence of the early-day activity can be seen in the piles of
boulders and old shafts five miles southwest of Franktown.

[23] Whitehill, H. R., Biennial Report of the State Mineralogist of the State of
Nevada for the years 1875 and 1876, pp. 63, 64.

Aside from their economic importance, the placer deposits are interesting from a geological viewpoint, as they occur in a Tertiary river channel. This ancient channel has been traced from a point near Incline on Lake Tahoe to a point a short distance east of the south shore of Washoe Lake, a distance of nearly eight miles. In places the river gravels are overlain by flows of rhyolite and andesite. The gravels consist of well-rounded pebbles and boulders, some of the latter being 30 inches in diameter. The gravels were derived from the bedrock mass of the Sierras and consist of quartzite, schist, granodiorite, and other rocks. The most productive part of the channel was in Little Valley, five miles southwest of Franktown. It is reported that placer gold having a value of $60,000 was taken from one pocket. Mining was hindered by the large amount of water and a number of faults, which displaced the gravels.

In 1935 two men were prospecting in the vicinity of the old workings on ground belonging to the Hobart Estate. Water for sluicing was obtained from Franktown Creek, which flows the year 'round.

PEAVINE DISTRICT

The Peavine District is in Washoe County, six miles by road northwest of Reno on the northern slope of Peavine Mountain. The lode deposits in the District were discovered in 1863, and it is interesting to note that the first rail shipment over the Sierras consisted of ore from this District shipped to Sacramento in 1866.

The only records of placer mining in the District are for the years 1876 to 1881 and the early nineties. The placer area is in a narrow ravine and is 1,500 feet long and from two to three feet deep, judging from the old workings. Several springs in the vicinity furnished water for sluicing, and when this supply failed, dry washing was employed. A number of rich boulders of ore were found while placer mining was being done. In recent years there has been no placer activity in the District.

OLINGHOUSE DISTRICT

The Olinghouse or White Horse District is on the east slope of the Wilcox Mountains of the Virginia Range in southeast Washoe County, about nine miles west of Wadsworth. The District was first prospected in 1860. Prior to 1900 the placer deposits in several ravines tributary to Olinghouse Canyon were of considerable importance and the output of placer gold, although not definitely known, was valued at many thousands of dollars. It is known that the residents of Wadsworth worked the placers for a number of years with pans and rockers and earned more than wages for their efforts. Some of these early placer miners treated gravel in the District that ran from $5 to $17 per cubic yard.

In 1935 W. W. Booth and Charles Barton were working together on placer ground owned by the Olinghouse Mining Company under a lease agreement whereby a royalty of 10 percent was paid on the gross returns. The work was confined to a small ravine half a mile long and 100 feet wide tributary to Olinghouse Canyon. This ravine had been worked in the early days by drift mining on bedrock. The depth of the placer material varies from 10 to 26 feet, averaging about 20 feet. The alluvium consists of subangular unassorted detrital

Figure 62. Home-made gravel-washing plant, Olinghouse District.

gravel, about 90 percent of which is less than one inch in size. The gold is both fine and coarse. The fineness averages 680. The largest nugget found in the District weighed one and one-half ounces. The placer gold probably was derived from the vein deposits on Green Mountain at the head of the ravine. The gravel is loosely cemented, and in places some clay is present. The best values are found in the five to six feet of gravel directly above bedrock.

The gravel is drift mined from a shaft 25 feet deep. Hoisting is done with a small bucket and hand windlass. After the gravel is hoisted it is dumped onto a 1-inch mesh screen placed over a two-wheel cart. The undersize drops into the cart and is hauled several hundred yards by automobile to a home-made washing plant, shown in figure 62, which has a capacity of ten cubic

yards of gravel in eight hours. The washing plant consists of
an amalgamator and sluice box. The amalgamator is made of
a 50-gallon gasoline drum, to the top of which an automobile
differential has been bolted. One axle extends into the barrel,
and to this axle six vertical rods are clamped. The rods reach
nearly to the bottom of the drum and serve as rakes to prevent
the gravel from packing. The other axle of the differential is
cut off and welded so that it will not turn. The differential gears
are driven by pulley attached to the free end of the drive shaft.
The gravel is shoveled into the amalgamator through an opening
cut into the side of the drum, and is discharged through another
opening in the side. About one tablespoon of quicksilver is
placed in the amalgamator at the beginning of a run and a tea-
spoonful added for every four cubic yards of gravel treated.
Water is fed to the amalgamator through a 1-inch pipe. Water
consumption is 200 gallons per cubic yard of gravel. A wind-
mill pump with 6-inch stroke is used for pumping water from
a nearby well 22 feet deep. The amalgamator and pump are
driven by a 1½-hp. Crowe gasoline engine. The cost of fuel
is 4 cents per cubic yard of gravel.

The sluice box below the amalgamator is 6 feet long, 12 inches
wide, 10 inches high, and is equipped with Hungarian riffles.
About 90 percent of the gold is caught in the amalgamator.

The cost of the entire washing plant, with second-hand equipment, is as follows:

Drum for amalgamator	$1
Differential	3
Three pulleys	15
1½-hp. gas engine	15
Windmill pump	12
Two drums for water tanks	2
Belts, piping, etc.	12
Total	$60

In 1935 the two lessees recovered an average of half an ounce
of gold per day. The value of the gravel treated was between
$1.50 and $2 per cubic yard.

Three dry-washing outfits also worked in the District in 1935.
Due to the moisture content of the gravel and its cemented character, dry-washing methods are not very successful in the District.

WHITE PINE COUNTY
BALD MOUNTAIN DISTRICT

The Bald Mountain or Joy District is in the northwestern part
of White Pine County on the western slope of the Ruby Range,

at an elevation of 7,400 feet.   The District was discovered in 1869, and placer gold was mined in the early days, principally by Chinese.   The amount of gold recovered was small.

The placer deposits are found in Water Canyon and tributary ravines.   The old workings in Water Canyon are outlined by pits and tailings extending up the canyon for about one mile west of the abandoned camp of Joy.   The richest ground was near the lower end of Water Canyon, where the channel is narrow and the gravel is about 10 feet deep.   At this point the pay streak on bedrock was 14 to 18 inches thick, overlain by 6 to 13 feet of overburden, which carried a little gold irregularly distributed. The gold found on the bedrock was fairly coarse.   Nuggets ranging in value from $2.50 to $10 were found.   Concentrate from the gravel contains scheelite.

The scarcity of water and the short mining season made placer mining difficult.   Small-scale methods were used by the early miners.   In recent years a number of attempts have been made to work placer ground in the District, but the results were not encouraging.

OSCEOLA DISTRICT

The Osceola District is on the west flank of the Snake Range in southwestern White Pine County, 40 miles southeast of Ely, Nevada.   The first lode discovery in the District was made by James Matteson in 1872, and considerable quartz mining was done for several years.   John Versan, an intelligent prospector, concluded that the gulches (one known as Grub and the other as Dry Gulch) would carry the gold that had been eroded from the quartz veins.   He organized a prospecting party and began to sink shafts.   His efforts were rewarded by finding very rich gravel at the junction of these two gulches, and in a few days following this discovery some 300 placer claims were located. The discovery of placer gold occurred in 1877 and marked the beginning of placer mining in the District, which has continued up to the present.

Prior to 1900 the output of placer gold is estimated to have been between $2,000,000 and $3,500,000, the first figure probably being more nearly correct.   From 1903 to 1933 the placer output was $51,531, as shown by the annual volumes of Mineral Resources of the United States.

The placers are distributed over a wide area, which includes Dry Gulch and Grub Gulch (also known as Wet Gulch) in the immediate vicinity of the town of Osceola, Weaver Creek, Mary Ann Canyon (Hogum placers), and the Summit placer diggings.

The main placer diggings are in Dry and Grub Gulches, which are the drainage channels for the area surrounding Osceola. The gulches are dry the greater part of the year.

The placer deposits in Dry and Grub Gulches vary from a thin covering at the upper ends to as much as 200 feet deep at the lower ends. Usually the gold is disseminated more or less through the gravels, but the principal pay streak is near bedrock. The bedrock is shale, limestone, and quartzite. The gravels contain but few large boulders. The gold is both coarse and fine, having a fineness of about 850.

It is reported that from 300 to 400 placer miners worked in these two gulches for several years following their discovery. These placers were worked by ground sluicing, rockers, and sluice boxes when water was available. According to an item in the White Pine News, dated October 6, 1888, dry washers were introduced into the District that year. Small-scale placer-mining operations continued into the early eighties. In the year 1878, on one of Versan's claims, a miner found a nugget weighing 24 pounds and valued at $3,600. This nugget was stolen and carried to the nearby camp of Ward, where it was melted into bars. The thief eventually repented of his deed and returned the bullion to the rightful owner. This nugget is perhaps the largest found in the State.

The important placers in Dry Gulch were consolidated in the early eighties and the Osceola Placer Mining Company was organized to work the deposits on a large scale by hydraulicking and sluicing. At a cost of $200,000 this company constructed a ditch and flume line having a total length of about 30 miles. One branch of the ditch ran on the east side of the range and brought in water from Baker and Lehman Creeks; the other branch was built on the west slope and brought water to the placer diggings from Williams, Pine, Shingle, Ridge, and Willard Creeks. The ditches were dug with scrapers and teams, and several small sections had to be blasted in rock. A White Pine News item states that up to August 31, 1886, 125,892 cubic yards of gravel had been hydraulicked with an average of 356 miner's inches of water available from six to eight hours per day. The yield from this amount of gravel was $29,715.23. The ditch line was begun in 1885, and the foregoing amount of water was obtained from only one branch. Later the other branch was completed, and it is reported that as much as 2,000 miner's inches of water was available from the two ditches. A view of a placer bank on the Hampton placer ground in Dry Gulch, which was hydraulicked in the early days, is shown in figure 63.

This bank is 46 feet high, and a 5 - cubic - yard channel sample taken by E. R. Wagner of Las Vegas, Nevada, from surface to the bottom of the bank is reported to have averaged $1.31 per cubic yard.

The Osceola Placer Mining Company continued hydraulicking until about 1900, when, on account of the loss in efficiency of the ditch from leaky flumes, light snowfall, stealing of the water from the ditch line, and other causes, work was discontinued.

In 1934 the Hampton placer, covering 417.74 acres of patented ground in Dry Gulch owned by W. N. Bowen, was bonded to the Wagner Gold Placer Company, Inc. Edgar R. Wagner,

Figure 63. Placer bank hydraulicked in the early days, Osceola District.

of Las Vegas, Nevada, is the principal stockholder. The treatment plant is shown in figure 64. Cost of the equipment was about $11,000.

The old workings in the upper portion of Dry Gulch was sampled by taking 174 cubic yards of gravel in 1-cubic-foot lots from several of the old shafts on the property. These samples ranged from 17 cents to $8.77 per cubic yard from surface to bedrock and averaged $1.32 per cubic yard. The shafts sampled ranged from 7 to 54 feet deep. The average depth of 124 holes was 26½ feet to bedrock. Sampling by drilling at Osceola is impracticable, as the quartzite boulders in the alluvium carry values in free gold up to $1.20 per ton and this gold would vitiate drill samples.

The gravel plant consists of two dragline scrapers, each driven by a 75-hp. Waukesha gasoline engine, and a washing plant made

by the Pioneer Gravel Equipment Manufacturing Company of
Milwaukee, Wis.

The gravel is hauled to the washing plant by one of the dragline outfits and is discharged into a hopper below the surface
of the ground.   Above the hopper is a grizzly made of ⅝ x 3-
inch strap iron with 3 - inch openings.   From the hopper, the
gravel is fed to an inclined conveyer belt 24 inches wide and
70 feet long, which discharges into the trommel.   The trommel
is 42 inches in diameter, 16 feet long, and is equipped with a
punched - plate screen having ⅝- and 1 - inch holes.   Oversize
from the trommel is discharged into a 21-cubic-yard steel bin
below, and from the bin it is discharged over a slide to the side
of the machine.   From the side of the machine the oversize is

Figure 64.   Placer plant of the Wagner Gold Placer Company, Dry Gulch,
Osceola District.

transported by the other dragline scraper to the waste dump
on the side of the hill.

The trommel undersize, constituting about 50 percent of the
material mined, is discharged by gravity to a sluice box 50 feet
long, 2 feet wide, 10 inches deep and sloping 1½ inches per foot.
Riffles in the sluice are made of 1 x 3-inch boards spaced 3 inches
apart and built in 5-foot sections.

The management states that the plant operated a few weeks in
1935 and during this period treated some 3,000 cubic yards of
gravel averaging 69½ cents per cubic yard.   The gravel treated
consisted mainly of tailings from former operations.   The plant
closed in October for the winter because of water shortage.

During 1934 and 1935 about 25 men were working placer

deposits in the District by small-scale hand methods. Most of these were working in the Hogum area on ground owned by T. B. Tilford. Royalty payments on Tilford ground vary from 25 to 35 percent of the gross returns.

The Hogum placers are three miles southwest of Osceola, on an alluvial fan that spreads out from Mary Ann Canyon. Pay gravel was found here in 1879, and the deposits have been worked intermittently since that time.

The placer deposits occur in channels buried under the detrital material of the alluvial fan. Usually they occur in a stratum overlying a false bedrock of cemented material. Frequently small potholes that carry high values are found in the false bedrock. It is believed that the gold was derived from the

Figure 65. Patented portable placer machine used in Hogum placer area, Osceola District.

quartzite strata exposed on the ridge above the canyon. The gold is generally fine.

The pay gravel is removed by drift mining and hoisted either by hand windlasses or small power hoists. In the spring of the year when water is available the gravel is sluiced. During the summer months hand-powered dry washers generally are used to recover the gold. In working with dry washers, sometimes the gravel has to be dried before it can be treated. Stoves made of sheet iron placed on rocks are employed for this purpose. Sagebrush is used for fuel.

During 1933 and 1934 William Trent worked in the Hogum area and recovered $7,500 in gold with a G. B. portable placer machine, shown in figure 65. This machine handled about two cubic yards per hour with a water consumption of 70 gallons per cubic yard. Water for placering was pumped to the ground through two miles of 2-inch pipeline with a Gould triplex pump, size 4 by 6 inches, belt driven by a 12 - hp. Fairbanks - Morse gasoline engine.

The Weaver Creek placer diggings are east of the divide, above the camp of Osceola. In the early days a considerable area at the upper end of the creek was worked by Chinese. In 1932 an attempt was made to work the gravels adjacent the creek by sluicing. The gravel was excavated and transported to the sluice by a small dragline scraper. Operations were hampered by large boulders and water on the bedrock. In addition, the placer operators became involved with ranchers below the placer workings over water rights. In 1935 there was very little placer activity in the vicinity of Weaver Creek.

The Summit Diggings are near the crest of the divide several miles above Osceola. In 1932 considerable excitement was aroused over the reported discovery of placer gold in the area. Many claims were staked out by miners from Ely, but, after a short period of prospecting, it was found that the ground did not come up to expectations, and the diggings were abandoned.

In 1935 some 4,000 acres of ground on the west slope of the Snake Range north of the Hogum placers were under option to the Osceola Gold Mining Corporation, Norman C. Stines, consulting engineer and director. The ground represents a consolidation of 19 separate placer holdings. During 1935 about 15 men were employed by the company in sampling. Prospecting shafts 4 x 5 feet in section were sunk on contract. The contract price for sinking shafts varied from $3 to $4 per linear foot, depending on character of material and the depth.

The pay gravel occurs in a series of channels on a false bedrock of caliche. The shallowest depth to the caliche is 30 feet. According to William Trent, who was instrumental in consolidating the various placer holdings, seven channels have been found, averaging about two miles in length and about 60 feet in width. Gold is found from the surface to the false bedrock, but the best values are in the five to six feet of gravel directly above the false bedrock. Most of the alluvium is wash gravel with a few boulders up to one foot in diameter. Some of the gravel is cemented, and blasting is necessary. The false bedrock varies in thickness from six inches to three feet. True bedrock on the alluvial fans has not been determined, but judging from the exposures at the east boundary of the placer ground it is probably quartzite to the north and limestone to the south. The gold is fairly coarse and it has a fineness of 844 to 857.

In Spring Valley, to the west of the placer ground, a large flow of subsurface water is available. A sump 15 by 15 feet in area and 20 feet deep was dug in the valley, and the water level in the sump is about 12 feet from the surface. A 250-gallon - per - minute pump was installed to test the amount of water available. Several weeks of steady pumping did not lower the water level. Near this sump is a drilled well, 750 feet deep, which has a flow of at least 400 gallons per minute by actual test. The water in this well rises to within about 12 feet from the surface.

In 1935 some preliminary sampling was done on placer ground on the east slope of the Snake Range about three and one-half miles west of Osceola. The Black Horse Gold Mining Company, financed by a group of business men from Houston, Tex., is planning to work the property.

The placer ground consists of the T. S. Mathis ranch of 160 acres, which is held under option, and 320 acres of located ground. This ground has never been worked. Three shafts, averaging 54 feet deep, were sunk on the property in 1935. Bedrock is said to be quartzite. The best values occur in the two feet of material above bedrock.

A flow of 65 miner's inches of water is available from seven springs on the property.

MINING ON RAILROAD GRANT LANDS

The Southern Pacific Railroad Company owns, or has legal rights in, many thousands of acres of desert land in Nevada. The question is often asked: "What right have I to a mineral discovery I have made upon railroad grant land?"

When railroad grant lands have been patented by the railroad company, no one may obtain title to any mineral thereon, or mine the same, without first purchasing or leasing the land from Southern Pacific Company, Land Department, San Francisco, Calif.

Therefore, the discoverer should first write to the Register and Receiver, United States Land Office, Carson City, Nevada, to ask if the land has or has not been patented. If the railroad grant land has not been patented, a discovery of valuable mineral thereon should be located and recorded in the usual manner, and the fact should be reported by letter to the Chief of the Field Division, United States General Land Office, Custom House, San Francisco, Calif. The General Land Office then will call the attention of Southern Pacific Company to the mineral character of the land, for lands of known mineral character were excluded from the railroad grant and may not be patented. The land in controversy is then re-examined by mining geologists employed by the railroad company, and also by mineral inspectors from the General Land Office. If the railroad company concedes the mineral character of the land, it is at once excluded from the railroad grant selection list. In case of disagreement between the engineers, a hearing is held before the receiver of the local Land Office at Carson City, Nevada. After the receiver has taken the sworn testimony of all witnesses each side cares to present, he will render a decision as to the mineral or non-mineral character of the land in controversy and eliminate from the railroad company's selection list all of the mineral land by legal subdivisions, after which the mining locations thereon may be patented.

Such a determination of the mineral or nonmineral character of railroad grant lands selected for patent is entirely without cost to the locator of the mining claim thereon.

SELECTED BIBLIOGRAPHY

This bibliography has been selected from textbooks, Government publications, and mining magazines containing information on placer mining.

The symbols following each reference have the following meanings:

*Available for consultation at large libraries.

†Obtainable from Superintendent of Documents, Washington, D. C.

‡Obtainable from Publications Section, U. S. Bureau of Mines, Washington, D. C.

§Obtainable from publisher.

M. & S. P.—Mining and Scientific Press.

E. & M. J.—Engineering and Mining Journal.

The references in this bibliography, as well as a number of others not listed, are available for consultation in the Mackay School of Mines Library, Reno, Nevada.

Allan, A. W.—Fundamentals of Amalgamation. E. & M. J., Vol. 116, Aug. 18, 1923, pp. 275–280.*
  The Berdan Pan for Amalgamation. E. & M. J., Vol. 106, Dec. 21, 1918, pp. 1075, 1076. Also, E. & M. J., Vol. 125, May 12, 1928.*

Andros, S. O.—Conservation of Water in Placer Prospecting. E. & M. J., Vol. 93, 1912, p. 1266. (Gives sketch of details of rocker.)*

Barbour, Percy E.—A California Dry Placer Dredge. E. & M. J., Vol. 94, 1912, p. 687.*

Bonery, Pierre—Study of Riffles for Hydraulicking. E. & M. J., May 24, 1913, pp. 1055–1060.*

Brooks, A. H.—Genesis and Classification of Placers. U. S. Geol. Survey Bull. 328, 1928.†

Birkenbine, John—Hydraulic Pumping Plant on the Snake River, Idaho. Trans. Am. Inst. Min. Eng., Vol. 30, 1900, p. 518.*

Bowie, A. J., Jr.—A Practical Treatise on Hydraulic Mining in California, 1885. Published by D. Van Nostrand, New York.*

Byron, E. L.—Special Machinery for Placer Mining. M. & S. P., Vol. 98, 1909, p. 128.

Campbell, D. F.—La Grange Hydraulic Mine. M. & S. P., Vol. 97, Oct. 10, 1908, pp. 491–493.*

Carter, T. L.—Gold Placers of Arizona—Dry Washings of Value. M. & S. P., Vol. 105, 1912, pp. 166–168.

Chapman, C. M.—Edison Dry Process for the Separation of Gold from Gravel. E. & M. J., Vol. 75, May 9, 1903, p. 713.*

Carey, E. E.—Working Dry Placers. American Metallurgical Soc., Vol. 1, 1911, p. 6.*

Dennis, F. J.—Modern Methods of Gravel Excavation; Steam Shovel and Drag-line Excavators. M. & S. P., Aug. 3, 1922, pp. 136–140.*

Dolbear, S. H.—Dry Placer Mining in California. E. & M. J., Vol. 89, 1910, p. 359.

Doolittle, J. E.—Gold Dredging in California. Bull. 36, California State Division of Mines, 1905.*

Endlich, F. M.—Mining in the Mojave Desert in California (Goler Dry Placers). E. & M. J., Vol. 62, Aug. 29, 1896, pp. 197, 198.*

Fairbanks, H. W.—Red Rock, Goler and Summit Mining Districts, Kern County, California. Twelfth Annual Report of California State Mineralogist, 1893–1894, pp. 456–458.*

Fansett, Geo. R.—Small-Scale Gold Placering. University of Arizona Bull., Vol. 3, No. 1, January 1, 1932, pp. 72–93.

Ficket, F. W.—Dry Placer Mining in the Quijotoa District (Arizona-Mexico). M. & S. P., Vol. 102, 1911, p. 291.

Ferguson, H. G.—The Round Mountain District, Nevada. U. S. Geol. Survey Bull. 825–I, 1921, pp. 383–406.

Franke, Herbert A.—Selected Bibliography on Placer Mining. Quarterly Report, California Division of Mines, April, 1932, pp. 219–224.*

Gardner, W. H.—Drilling for Placer Gold. Published by Keystone Drilling Company, Beaver Falls, Pa., 194 pp.

Gross, J.—Comparison of Methods of Gold Recovery from Black Sand. U. S. Bureau of Mines, Report of Investigations, No. 2192, 1920.*
Recovery of Gold from Black Sand Tailings. U. S. Bureau of Mines, Report of Investigations, No. 2170, 1920.*

Haley, Chas. S.—Gold Placers of California. Bull. 92, California State Division of Mines, 1923.*
Dry Placers of Southern California. (Abstract) E. & M. J.–Press, Vol. 1. Dec., 1922, pp. 235, 236.*
Elevating Ten-Cent Gravel at a Profit. M. & S. P., Vol. 104, 1912, pp. 530–532.*

Hartley, C.—Opportunity in Placer Mining. E. & M. J., Vol. 99, 1915, pp. 185–208.*
The Recovery of Fine Gold and Black Sand. E. & M. J., Vol. 97, 1914, pp. 843, 844.*

Hazard, F. H.—The Saving of Fine Placer Gold. E. & M. J., Vol. 92, Aug. 26, 1911, pp. 394–396.*

Heikes, V. C., and Yale, Chas. G.—Dry Placers in Arizona, California, New Mexico, and Nevada. Mineral Resources, U. S. Geol. Survey, 1912, pp. 254–263.*

Hill, J. M.—Notes on the Fine Gold of Snake River, Idaho. Bull. 620–L, U. S. Geol. Survey, 1916.*
Notes on Placer Deposits of Greaterville, Arizona. U. S. Geol. Survey Bull. 430, 1910, pp. 11–22.†

Horner, R. R.—Notes on Black Sand Deposits of Southern Oregon and Northern California. Technical Paper 196, U. S. Bureau of Mines, 1918, 42 pp.‡

Hutchins, J. P.—The Nomenclature of Modern Placer Mining. E. & M. J., Vol. 84, Aug. 17, 1907, pp. 293–296.*
The Essential Data of Placer Investigations. E. & M. J., Vol. 84, 1907, pp. 339–342, 385–387.*

Ingersoll, Guy E.—Hand Methods of Placer Mining and Placer Mining Districts of Washington and Oregon. Bulletin of State College of Washington, March, 1932.*

Jackson, C. F., and Knabel, J. B.—Small-Scale Placer Mining Methods. Information Circular 611, U. S. Bureau of Mines, 1932, 17 pp.*

Janin, Charles—Placer Mining Methods and Operating Costs. Chapter of Bulletin 121, U. S. Bureau of Mines, 1916.*
Gold Dredging in the United States. U. S. Bureau of Mines Bull. 127, 1918, 226 pp.

Jardine, J. B., Jr.—Dry Concentration of Placer Gold. American Metallurgical Society, Vol. 1, 1911, p. 21.

Keizer, W. G.—Dry Placer Mining on a Large Scale. (Arizona) Mining and Engineering World, Vol. 44, 1916, pp. 999, 1000.*

Kirkpatrick, T. S. G.—The Hydraulic Gold Miner's Manual. Second Edition, 1897. Published by E. and F. N. Spon, London and New York.

Knox, H. B., and Haley, C. S.—The Mining of Alluvial Deposits. The Mining Journal, London, Vol. 12, No. 2, p. 89; Vol. 12, No. 3, p. 153; Vol. 12, No. 4, p. 211, 1915.*

Laizure, C. McK.—Some Special Methods and Machines for Recovery of Gold and Platinum in Placer Deposits. Quarterly Report, California Division of Mines, January, 1929, pp. 94–135.
Elementary Placer Mining Methods and Gold Saving Devices. Quarterly Report, California Division of Mines, April, 1932, pp. 112–204.*

Leaver, E. S.—Recovery of Fine Gold by Amalgamation. E. & M. J., Vol. 127, April 13, 1929, pp. 601, 602.*

Lindgren, Waldemar—The Tertiary Gravels of the Sierra Nevada of California. Professional Paper 73, U. S. Geol. Survey, 1911.*

Mather, Henry A.—The Problem of the Dry Placers. E. & M. J., Vol. 76, Aug. 29, 1903, pp. 314, 315.*

Merrill, F. J. H.—Dry Concentration of Placer Gold. M. & S. P., Vol. 105, 1912, pp. 50–52.*

Packard, George A.—Round Mountain Camp, Nevada. E. & M. J., Vol. 83, 1907, pp. 150, 151.*
Round Mountain, Nevada. M. & S. P., Vol. 96, 1908, pp. 807–809.

Peele, Robert—Mining Engineer's Handbook. Second Edition, 1927 (two volumes). Published by John Wiley & Sons, Inc., New York.§

Peterson, G. M.—Dry Placer Machines. M. & S. P., Vol. 103, 1910, p. 639.*

Perret, L. A.—Gold in Black Sand (Berdan Pan). M. & S. P., Vol. 123, 1921, pp. 423, 424.*

Plummer, Wm. L.—Successful Dry Placer Operations at Plumosa, Arizona. Engineering and Mining World, Vol. 45, 1916, pp. 1–3.*

Purington, C. W.—Methods and Costs of Gravel and Placer Mining in Alaska. Bulletin 263, U. S. Geol. Survey, 1905.*

Radford, W. H.—Notes on Hydraulic Mining of Low Grade Gravel. Vol. 31, Transactions of American Institute of Mining Engineers, 1907, p. 918.*

Richards, J. V.—Dry Washing for Placer Gold in Sonora, Mexico. Transactions of American Institute of Mining Engineers, Vol. 41, 1911, pp. 797–802.

Rickard, T. A.—The Alluvial Deposits of Western Australia. Transactions of American Institute of Mining Engineers, Vol. 28, 1898, pp. 490–537.
The Alluvial Deposits of Western Australia. M. & S. P., 1899, Vol. 78, pp. 238, 266, 292, 293, 349, 350, 330, 331.

Sampson, J. R.—Placers of Southern California. Quarterly Report, California Division of Mines, April, 1932, pp. 225–255.*

Sperry, E. A.—Recovery of Flour Gold from River Sand. E. & M. J., Vol. 128, Oct. 12, 1929, pp. 581, 582.*

Staley, W. W.—Elementary Methods of Placer Mining. Pamphlet No. 35, Idaho Bureau of Mines and Geology, 1931.*

Von Bernewitz, M. W.—Handbook for Prospectors. Second Edition, 1931. Published by McGraw-Hill Book Company, Inc., New York.§

Washburn, Wm. H.—Method of Saving Fine Gold of the Snake River, Idaho. M. & S. P., Vol. 83, 1901, pp. 45, 46.*

Wilson, Eldred D., and Tenny, J. B.—Arizona Gold Placers and Placering. University of Arizona Bulletin, Vol. 3, No. 1, Jan. 1, 1932, pp. 9–72.

Winston, W. B., and Janin, Charles—Gold Dredging in California. Bulletin 57, California State Division of Mines, 1910.*

Wolcott, G. E.—Mining and Milling at Rawhide, Nevada (Dry Placer Mining). E. & M. J., Vol. 87, 1909, p. 346.*

Wright, W. H.—Laying out Lines and Computing Yardage. E. & M. J., 1914.

Yale, C. G.—Dry Placers of California. Mineral Resources of the United States, U. S. Geol. Survey, Part 1, 1912, pp. 262, 263.*

Yeatman, J. A.—The Pump in Placer Mining. M. & S. P., Vol. 88, 1904, p. 226.

Young, Geo. J.—Elements of Mining. Second Edition, 1923. Published by
    McGraw-Hill Book Company, New York.§
    Drift Mining at Callecito, Calaveras County, California. E. & M. J., Vol.
    128, Sept. 7, 1929, pp. 394–397.*
Anonymous—Ingenious Placer Operation near Manhattan, Nevada. Engineer-
    ing and Mining World, Vol. 39, 1913, p. 209.*
    The Altar Gold Placer Fields of Sonora, Mexico (Describes the Quenner
    Pulverizer). E. & M. J., Vol. 90, 1910, pp. 651–653.*
    Placer Mining Methods. Reports of Investigations, No. 2315, U. S. Bureau
    of Mines, January, 1922, 4 pp.
    Dry Blowing Gold-Bearing Gravel. M. & S. P., Vol. 89, 1904, p. 328.

NEVADA STATE BUREAU OF MINES AND MACKAY SCHOOL OF MINES PUBLICATIONS

An asterisk (*) indicates the bulletin is out of print. Prepayment is required and may be made by postal or express money order, payable to the Nevada State Bureau of Mines, or in cash, exact amount, at sender's risk. Postage stamps are not accepted in payment for publications.

PRICE

Preliminary Report on the Building Stones of Nevada, by John A. Reid, 1904*..

The Ventilating System at the Comstock Mines, Nevada, by George J. Young, 1909*..

Fires in Metalliferous Mines, by George J. Young, 1912*......

Manganese, by Walter S. Palmer, 1918*................................

Vol. 1, No. 1. Mineral Resources of Southern Nevada, by Jay A. Carpenter, 1929.. $0.20

Vol. 22, No. 1. Identification of Nevada's Common Minerals, with Notes on Their Occurrence and Use, by Oliver R. Grawe, 1928—

With mineral identification chart............................. .50

Without mineral identification chart....................... .25

Vol. 22, No. 2. Dumortierite, by Mackay School of Mines Staff, 1928*........ ....

Vol. 24, No. 4. The Underground Geology of the Western Part of the Tonopah Mining District, Nevada, by Thomas B. Nolan, 1930* ....

Vol. 25, No. 3. Notes on Ore Deposits at Cave Valley, Patterson District, Lincoln County, Nevada, by F. C. Schrader, 1931......... .20

Vol. 25, No. 4. A Preliminary Survey of the Scossa District, Pershing County, Nevada, by J C. Jones, A. M. Smith, and Carl Stoddard, 1931*..

Vol. 25, No. 5. Ore Deposits of the Gold Circle Mining District, Elko County, Nevada, by Edward H. Rott, Jr., 1931.............. .20

Vol. 25, No. 6. Bedded Deposits of Manganese Oxides near Las Vegas, Nevada, by D. F. Hewett and B. N. Webber, 1931...... .20

Vol. 25, No. 7. Cherry Creek (Egan Canyon) District, White Pine County, Nevada, by F. C. Schrader, 1931..............................

The Spruce Mountain District in Elko County, Nevada, by F. C. Schrader, 1931.............................. } .25

Vol. 26, No. 5. The Mines and Mills of Silver City, Nevada, by A. M. Smith, 1932.. .20

Mining Districts and Mineral Resources of Nevada, by Francis Church Lincoln, 1923.............. $1.50

Vol. 26, No. 6. Metal and Nonmetal Occurrences in Nevada, 1932... .50 } 1.50

Vol. 26, No. 7. Nonmetallic Minerals in Nevada, by J. A. Fulton and A. M. Smith, 1932.. .10

Vol. 26, No. 8. Placer Mining in Nevada, by Alfred Merritt Smith and W. O. Vanderburg, 1932*...................................... .25

⑳

www.ingramcontent.com/pod-product-compliance
Lightning Source LLC
Chambersburg PA
CBHW022056210326
41519CB00054B/540